CONTROLLED ENVIRONMENT GUIDELINES FOR PLANT RESEARCH

ACADEMIC PRESS RAPID MANUSCRIPT REPRODUCTION

PROCEEDINGS OF THE CONTROLLED ENVIRONMENTS
WORKING CONFERENCE HELD AT MADISON, WISCONSIN
MARCH 12–14, 1979

CONTROLLED ENVIRONMENT GUIDELINES FOR PLANT RESEARCH

Edited by

T. W. TIBBITTS
Department of Horticulture
University of Wisconsin
Madison, Wisconsin

and

T. T. KOZLOWSKI
The Biotron
University of Wisconsin
Madison, Wisconsin

ACADEMIC PRESS
A Subsidiary of Harcourt Brace Jovanovich, Publishers
1979   New York   London   Toronto   Sydney   San Francisco

ACADEMIC PRESS, INC.
111 Fifth Avenue, New York, New York 10003

United Kingdom Edition published by
ACADEMIC PRESS, INC. (LONDON) LTD.
24/28 Oval Road, London NW1 7DX

Library of Congress Cataloging in Publication Data

Controlled Environments Working Conference, Madison,
     Wis. , 1979.
     Controlled environment guidelines for plant
research.

     Includes index.
     1. Growth cabinets and rooms——Environmental
engineering——Congresses. 2. Phytotrons——Environ-
mental engineering——Congresses. 3. Botanical
research——Congresses. I. Tibbitts, T. W.
II. Kozlowski, Theodore Thomas.
III. Title.
QK715.5.C66 1979   581'.072'4   79—23521
ISBN 0—12—690950—4

CONTENTS

CONTENTS

CONTRIBUTORS

Numbers in parentheses indicate the pages on which the authors' contributions begin.

WADE L. BERRY (369), Laboratory of Nuclear Medicine and Radiation Biology, University of California, Los Angeles, California

E. D. BICKFORD (47), Duro-Test Corporation, North Bergen, New Jersey

G. S. CAMPBELL (301), Department of Agronomy and Soils, Washington State University, Pullman, Washington

R. B. CURRY (1), Department of Agricultural Engineering, Ohio Agricultural Research and Development Center, Wooster, Ohio

ROBERT J. DOWNS (29), Phytotron, North Carolina State University, Raleigh, North Carolina

MURRAY E. DUYSEN (381), Department of Botany, North Dakota State University, Fargo, North Dakota

JOHN S. FORRESTER (177), Scientific Systems Corporation, Baton Rouge, Louisiana

LAWRENCE J. GILES (229), Phytotron, Duke University, Durham, North Carolina

P. ALLEN HAMMER (343), Department of Horticulture, Purdue University, West Lafayette, Indiana

HENRY HELLMERS (229), Phytotron, Duke University, Durham, North Carolina

GLENN J. HOFFMAN (141), United States Salinity Laboratory, Riverside, California

MERRILL R. KAUFMANN (291), Rocky Mountain Forest and Range Experiment Station, USDA–Forest Service, Fort Collins, Colorado

HERSCHEL H. KLUETER (235), Agricultural Equipment Laboratory, USDA–SEA, Agricultural Research, Beltsville, Maryland

HENRY J. KOSTKOWSKI (331), National Bureau of Standards, Washington, D.C.

T. T. KOZLOWSKI (1), Biotron, University of Wisconsin, Madison, Wisconsin

PAUL J. KRAMER (391), Department of Botany, Duke University, Durham, North Carolina

DONALD T. KRIZEK (241), Plant Stress Laboratory, USDA–SEA, Agricultural Research, Beltsville, Maryland

K. J. McCREE (11), Department of Soil and Crop Sciences, Texas A & M University, College Station, Texas

J. CRAIG McFARLANE (55), Monitoring and Support Laboratory, Environmental Protection Agency, Las Vegas, Nevada

J. E. PALLAS, JR. (207), Southern Piedmont Center, USDA–SEA, Watkinsville, Georgia

LAWRENCE R. PARSONS (135), Department of Horticultural Science and Landscape Architecture, University of Minnesota, St. Paul, Minnesota

R. P. PRINCE (1), Department of Agricultural Engineering, University of Connecticut, Storrs, Connecticut

S. L. RAWLINS (271), United States Salinity Laboratory, Riverside, California

FRANK B. SALISBURY (75), Plant Science Department, Utah State University, Logan, Utah

R. P. SEARLS (131), Sherer Environmental Division, Kysor Industrial Corporation, Marshall, Michigan

L. A. SPOMER (193), Department of Horticulture, University of Illinois, Urbana, Illinois

C. B. TANNER (117), Department of Soil Science, University of Wisconsin, Madison, Wisconsin

G. W. THURTELL (173), Soils Department, University of Guelph, Guelph, Ontario, Canada

T. W. TIBBITTS (1), Department of Horticulture, University of Wisconsin, Madison, Wisconsin

ALBERT ULRICH (369), Department of Soils and Plant Nutrition, University of California, Berkeley, California

N. SCOTT URQUHART (343), Department of Experimental Statistics, New Mexico State University, Las Cruces, New Mexico

C. H. M. van BAVEL (323), Department of Soil and Crop Sciences, Texas A & M University, College Station, Texas

OTHER PARTICIPANTS

G. R. AMBURN, Laboratory of Hygiene, University of Wisconsin, Madison, Wisconsin

D. L. ANDERSON, Department of Soil Science, University of Wisconsin, Madison, Wisconsin

L. ANDERSON, Biotron, University of Wisconsin, Madison, Wisconsin

W. A. BAILEY, USDA–SEA, Beltsville, Maryland

M. BATES, General Mills, Inc., Minneapolis, Minnesota

C. BAUM, Biotron, University of Wisconsin, Madison, Wisconsin

W. BIGGS, LiCor, Lambda Instrument Co., Lincoln, Nebraska

W. BLAND, Department of Horticulture, Pennsylvania State University, University Park, Pennsylvania

H. BORG, Department of Soil Science, University of Wisconsin, Madison, Wisconsin

P. BROWN, Department of Soil Science, University of Wisconsin, Madison, Wisconsin

B. BUGBEE, Department of Horticulture, Pennsylvania State University, University Park, Pennsylvania

R. CARLSON, University of Illinois, 289 Morrill Hall, Urbana, Illinois

P. J. COYNE, Lawrence Livermore Laboratory, Livermore, California

C. O. CRAMER, Department of Agricultural Engineering, University of Wisconsin, Madison, Wisconsin

B. F. DETROY, Department of Agricultural Engineering, University of Wisconsin, Madison, Wisconsin

R. DICKSON, Forestry Sciences Laboratory, USDA–Forest Service, Rhinelander, Wisconsin

R. DOERING, Forest Products Laboratory, USDA–Forest Service, Madison, Wisconsin

A. ELLIS, Department of Horticulture, University of Wisconsin, Madison, Wisconsin

M. FLUCHERE, General Foods Corporation, Tarrytown, New York

A. B. FRANK, USDA–SEA, Agricultural Research, Mandan, North Dakota

H. FRANK, Department of Horticulture, University of Wisconsin, Madison, Wisconsin

W. R. GARDNER, Department of Soil Science, University of Wisconsin, Madison, Wisconsin

R. GLADON, Department of Horticulture, Iowa State University, Ames, Iowa

M. GUERRA, Environmental Growth Chambers, Chagrin Falls, Ohio

T. HAZEN, Agriculture and Home Economics Experiment Station, Iowa State University, Ames, Iowa

M. J. JAFFE, Department of Botany, Ohio University, Athens, Ohio

T. KIMMERER, Department of Forestry, University of Wisconsin, Madison, Wisconsin

E. KLADIVKO, Department of Soil Science, University of Wisconsin, Madison, Wisconsin

J. KOBRIGER, Department of Horticulture, University of Wisconsin, Madison, Wisconsin

R. KOCHHANN, Department of Soil Science, University of Wisconsin, Madison, Wisconsin

D. KOLLER, Department of Agricultural Botany, Hebrew University, Rehovot, Israel

A. LANG, Department of Horticulture, University of Wisconsin, Madison, Wisconsin

R. W. LANGHANS, Department of Floriculture and Ornamental Horticulture, Cornell University, Ithaca, New York

W. MAHON, Environmental Growth Chambers, Chagrin Falls, Ohio

J. MARANVILLE, Department of Agronomy, University of Nebraska, Lincoln, Nebraska

D. McCARTY, Department of Agronomy, University of Wisconsin, Madison, Wisconsin

B. H. McCOWN, Department of Horticulture, University of Wisconsin, Madison, Wisconsin

D. D. McCOWN, Department of Soil Science, University of Wisconsin, Madison, Wisconsin

G. McKEE, Department of Horticulture, Pennsylvania State University, University Park, Pennsylvania

E. MILLER, Department of Physics, University of Wisconsin, Madison, Wisconsin

N. D. NELSON, Forestry Sciences Laboratory, USDA–Forest Service, Rhinelander, Wisconsin

R. NORBY, Department of Forestry, University of Wisconsin, Madison, Wisconsin

R. NORTON, Northwest Washington Research Unit, Washington State University, Mount Vernon, Washington

G. NUTTER, Instrumentation Systems Center, University of Wisconsin, Madison, Wisconsin

C. OLDENBURG, Department of Horticulture and Landscape Architecture, Washington State University, Pullman, Washington

D. OLSZYK, Department of Horticulture, University of Wisconsin, Madison, Wisconsin

D. ORMROD, Department of Horticultural Science, University of Guelph, Guelph, Ontario, Canada

H. PHIPPS, Forestry Sciences Laboratory, USDA–Forest Service, Rhinelander, Wisconsin

C. A. PORTER, Monsanto Company, St. Louis, Missouri

B. POST, USDA–SEA, Washington, D.C.

D. RAPER, Department of Soil Science, North Carolina State University, Raleigh, North Carolina

M. READ, General Mills, Inc., Minneapolis, Minnesota

A. RULE, Environmental Growth Chambers, Chagrin Falls, Ohio

J. C. SAGER, Smithsonian Institution, Rockville, Maryland

L. SCHRADER, Department of Agronomy, University of Wisconsin, Madison, Wisconsin

V. SCHRODT, Monsanto Company, St. Louis, Missouri

J. SCHWARZ, Biotron, University of Wisconsin, Madison, Wisconsin

H. A. SENN, 2815 Admirals Road, Victoria, British Columbia, Canada

F. SKOOG, Department of Botany, University of Wisconsin, Madison, Wisconsin

R. H. TAYLOR, Controlled Environments Ltd. Winnipeg, Manitoba, Canada

F. W. TELEWSKI, Department of Botany, Ohio University, Athens, Ohio

T. TISCHNER, Agricultural Research Institute, Hungarian Academy of Sciences, Martonvasar, Hungary

L. E. TOWILL, Department of Horticulture, University of Wisconsin, Madison, Wisconsin

J. F. van STADEN, Agricultural Engineering Division, Silverton, South Africa

F. C. VOJTIK, Department of Plant Pathology, University of Wisconsin, Madison, Wisconsin

R. B. WALKER, Department of Botany, University of Washington, Seattle, Washington

D. L. WALTERS, Sherer Environmental Division, Kysor Industrial Corp., Marshall, Michigan

G. WANEK, Biotron, University of Wisconsin, Madison, Wisconsin

L. WARMANEN, Biotron, University of Wisconsin, Madison, Wisconsin

J. A. WEBER, University of Michigan, Biological Station, Ann Arbor, Michigan

F. WENT, Desert Research Institute, Reno, Nevada

S. WIETHOLTER, Department of Soil Science, University of Wisconsin, Madison, Wisconsin

D. WHEELER, USDA, Forest Service, Forest Products Laboratory, Madison, Wisconsin

J. WURM, LiCor, Lambda Instrument Co., Lincoln, Nebraska

PREFACE

This volume brings together information presented at the Controlled Environments Working Conference held in Madison, Wisconsin, March 12-14, 1979. The conference arose from the realization that progress in understanding the effects of environmental factors on plant growth has been greatly impeded because of the lack of uniformity in measuring and reporting environmental conditions in controlled environment facilities. This lack has made it nearly impossible for different investigators to compare their research findings. Furthermore, the lack of guidelines has led to the production of a variety of monitoring instruments with variable specifications. Publication of research papers has often been delayed because opinions of reviewers and editors have differed widely on what constitutes appropriate environmental measurements and reporting units.

The guidelines proposed include recommendations for regulating and measuring such environmental factors as radiation, temperature, carbon dioxide, atmospheric moisture, soil moisture, and air movement in controlled environment facilities. They suggest how measurements can be made accurately and in ways that can be repeated by other investigators.

The book is intended for biologists and engineers using controlled environments with a view toward ensuring precise and reproducible research. It should also be useful for investigators undertaking environmental measurement and control in greenhouses and in the field.

The papers were presented by invited plant physiologists, physicists, and engineers of demonstrated competence. We thank each contributor for his scholarly contribution as well as for his patience and cooperation during the production phases of this volume.

The conference was initiated by the North Central Regional Growth Chamber Committee of the U.S. Department of Agriculture–Science Education Agency. This conference was cosponsored by the conference by the University of Wisconsin Biotron, the Growth Chamber Working Group of the American Society for Horticultural Sciences, and the Committee on Environment and Plant Structures of the American Society of Agricultural Engineers. Representing these organizations on the conference planning committee were T.

W. Tibbitts (Chairman), T. T. Kozlowski, J. C. McFarlane, and R. L. Prince, respectively. Members of the local arrangements committee included L. C. Anderson, C. O. Cramer, T. T. Kozlowski, B. H. McCown, L. E. Schrader, C. B. Tanner, and T. W. Tibbitts. The help of Barbara A. Jungheim, Gay W. Stauter, and Susan Higgins in typing the manuscript is appreciated.

Financial support for the conference was provided by the Environmental Biology Division of the National Science Foundation, the U.S. Department of Agriculture, General Mills, the University of Wisconsin Graduate School, and the University of Wisconsin College of Agricultural and Life Sciences.

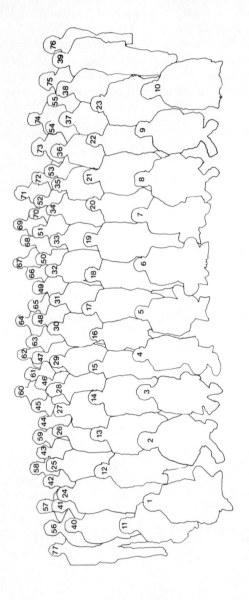

NUMBERS AND NAMES OF PEOPLE IN THE PHOTOGRAPH

1 J. Weber; 2 N. Nelson; 3 J. Wurm; 4 H. Kostkowski; 5 J. Forrester; 6 J. Van Staden; 7 D. Olszyk; 8 H. Phipps; 9 B. McCown; 10 A. Lang; 11 B. Post; 12 T. Tischner; 13 H. Hellmers; 14 F. Went; 15 T. Kozlowski; 16 H. Senn; 17 R. Prince; 18 T.Tibbitts; 19 R. Curry; 20 T. Hazen; 21 C. McFarlane; 22 P. Kramer; 23 D. Koller; 24 W. Berry; 25 L. Parsons; 26 R. Taylor; 27 L. Warmanen; 28 M. Bates; 29 W. Mahon; 30 J. Sager; 31 M. Read; 32 M. Kaufmann; 33 S. Rawlins; 34 G. Thurtell; 35 G. McKee; 36 P. Coyne; 37 P. Hammer; 38 R. Norton; 39 E. Bickford; 40 J. Maranville; 41 J. Schwarz; 42 D. Paige; 43 R. Downs; 44 R. Dickson; 45 M. Fluchere; 46 R. Searls; 47 A. Frank; 48 G. Hoffman; 49 L. Spomer; 50 M. Duysen; 51 M. Jaffe; 52 K. McCree; 53 G. Campbell; 54 R. Langhans; 55 D. Krizek; 56 L. Towill; 57 C. Baum; 58 D. Walters; 59 R. Carlson; 60 F. Salisbury; 61 D. Raper; 62 C. Tanner; 63 D.Ormrod; 64 L. Anderson; 65 T. Kimmerer; 66 W. Bailey; 67 B. Detroy; 68 C. Porter; 69 H. Klueter; 70 V. Schrodt; 71 F. Telewski; 72 F. Vojtik; 73 R. Gladon; 74 J. Pallas; 75 B. Bugbee; 76 R. Norby; 77 R. Walker

PARTICIPANTS

INTRODUCTION

R. B. Curry

Department of Agricultural Engineering
Ohio Agricultural Research and Development Center
Wooster, Ohio

T. T. Kozlowski

Biotron
University of Wisconsin
Madison, Wisconsin

R. P. Prince

Department of Agricultural Engineering
University of Connecticut
Stoors, Connecticut

T. W. Tibbitts

Department of Horticulture
University of Wisconsin
Madison, Wisconsin

The development of equipment for growing plants in controlled, reproducible environments is widely recognized as an extremely important contribution to research in plant science. Controlled environments have been advantageously used not only for producing uniform plants for biochemical studies but also for a variety of studies including those on: (1) effects of environmental factors on growth, (2) adaptation and acclimation, (3) screening plant material for special phenological characteristics, (4) interactions

1

between plants and other organisms, (5) effects of various bio-
cides under different environmental conditions, and (6) determination of environmental factors that are likely to be significant in influencing growth in the field (Kramer, Slatyer, and
Hellmers, 1972). Unfortunately, however, progress in extending
knowledge has been significantly slowed because of the lack of
accepted standardized procedures for characterizing and measuring
both above-ground and below-ground factors of the environment
within growth chambers.

Detailed and complete measurements are needed because of the
large variations in environmental conditions in different
laboratories and chambers, even though attempts are made to
maintain similar controls. These differences occur because of
variations in the direction of air flow, reflectivity of surfaces,
carbon dioxide concentrations of makeup air, size of chambers,
and various other reasons.

Precision could be greatly improved with guidelines and
uniformity in relation to: (1) sensors (types, accuracy and
precision, calibration procedures), (2) measurement (locus of
measurement for different species and stages of plant growth,
time of taking measurements, spacing of measurements, etc.),
and (3) reporting (appropriate units, conciseness and format,
etc.).

With few notable exceptions, there has been little organized
effort by plant scientists to standardize measurement and reporting of environmental parameters in growth chambers. One of the
earliest efforts by plant scientists to provide guidelines for
measuring and reporting was that of the Committee of Plant
Irradiation of the Netherlands Stichting voor Verlichtingskunde
in 1953. This committee recommended that irradiance for purposes
of practical plant irradiation be characterized with the aid of
several different spectral regions. The committee recognized
certain limitations in these broad categories and suggested

that they would be useful primarily for practical agriculture
and not for fundamental photobiological research. The committee
members may have been overly cautious in encouraging use of
these irradiance groupings by research scientists, but certainly
the lack of reasonably priced instrumentation for measurement
of irradiance within each spectral region has limited their
application in plant research.

Over the years there have been several conferences in which
measurement, instrumentation, and procedures for controlled
environments have been discussed and particular procedures
encouraged (Chouard and de Bildering, 1972; Evans, 1963; Hudson,
1957; Rees et al., 1972; White, 1963). One of the most detailed
discussions occurred in 1962 at conferences in Australia
associated with the opening of the Canberra Phytotron. In
association with a conference at Durham, North Carolina, convened
in 1972, there was a roundtable discussion on international
standards for measurement and an ad hoc committee was appointed
and charged with developing standards for:

1. Definition of the environmental factors to measure
    or describe.

2. Methods of making measurements.

3. Units in which the measurements should be expressed.

4. Methods of implementing standards.

Although this committee has not made specific recommendations,
its appointment does demonstrate the need for guidelines that
investigators would like to see put into practice.

In 1972, a committee from the American Society for Horticultural Science developed reporting guidelines (ASHS Committee
on Growth Chamber Environments, 1972), to indicate to investigators and journal editors what environmental parameters should be
measured and recommended procedures for reporting data. A sample
reporting paragraph was provided with the guidelines to assist
researchers in presenting information clearly and concisely in

research papers.   These guidelines were revised and republished
in 1977 (ASHS Committee on Growth Chamber Environment, 1977).
They provide strong encouragement to fully document the experimental environment.   There was no attempt in these guidelines to
require the recommended procedures, but they have been generally
accepted by the American Society of Horticultural Science;
consequently, there has been significant improvement in reporting
of environmental research.

The impetus for the set of proposed guidelines, presented
for discussion at this conference (Table 1), grew out of
discussions in a North Central Regional Committee of USDA/SEA
on Growth Chamber Use.   This committee, from its inception in
1976, had a common concern for standards in plant growth studies,
and particularly for the environment in which research was conducted in growth chambers.   Two important questions directed the
committee's concern: (1) how can the investigator be certain
that successive studies in his chambers can be compared with
respect to environmental conditions, and (2) how does the investigator compare with confidence the various studies reported in
the literature.

For the establishment of guidelines the committee decided to:
(1) develop a preliminary set of guidelines, (2) organize a
special conference to critique these guidelines, and (3) work
with the American Society of Agricultural Engineers, an organization that already had a procedure for developing, publishing,
and maintaining guidelines and standards.

The guidelines discussed at this conference have gone through
various critiques and editing processes.   Members of the North
Central Regional Committee of USDA/SEA evaluated them 2 or 3
times.   At least four drafts were prepared and reviewed by a
Committee of the American Society of Agricultural Engineers
(SE-303 Committee).   These reviews were guided by the need to
completely describe the environment and to have realistic requirements that can be met by all researchers.   Thus the emphasis by

TABLE 1. Proposed Guidelines for Measuring and Reporting the Environment for Plant Studies[a]

Parameter	Units	Measurements		
		Where to take	When to take	What to report
Radiation:				
PAR (Photosynthetically active radiation)				
a) Quantum flux density 400–700 nm with cosine correction	$\mu E\ s^{-1}\ m^{-2}$	At top of plant canopy. Obtain average over plant growing area.	At start and finish of each study and biweekly if studies extend beyond 14 days.	Average over containers at start of study. Decrease or fluctuation from average over course of the study.
or				
b) Irradiance 400–700 nm with cosine correction	$W\ m^{-2}$	(Same as quantum flux density)	(Same as quantum flux density)	(Same as quantum flux density)
Total irradiance 280–2800 nm with cosine correction	$W\ m^{-2}$	(Same as quantum flux density)	At start of each study.	Average over containers at start of study.
Spectral irradiance 250–850 nm in <20 nm bandwidths with cosine correction	$\mu E\ s^{-1}\ m^{-2}\ nm^{-1}$ or $W\ m^{-2}\ nm^{-1}$	At top of plant canopy in center of growing area.	At start of each study.	Graph of irradiance for separate wavebands at start of study.
Photometric[b] 380–780 nm with cosine correction	klx	(Same as quantum flux density)	At start of each study.	(Same as total irradiance)

[a]Proposed by the USDA/SEA North Central Regional Committee on Growth Chamber Use.

[b]Report with PAR reading for historical comparison.

Parameter	Units	Measurements		What to report
		Where to take	When to take	
Temperature:				
Air				
Shielded and aspirated (>3 m sec⁻¹) device	$°C$	At top of plant canopy. Obtain average over plant growing area.	Hourly over the period of the study. (Continuous measurement advisable)	Average of hourly average values for the light and dark periods of the study with range of variation over the growing area.
Soil and liquid	$°C$	In center of representative container.	Hourly during the first 24 hr of the study. (Hourly measurements over the period of the study, advisable)	Average of hourly average values for the light and dark periods for the first day or over entire period of the study if taken.
Atmospheric moisture:				
Shielded and aspirated (>5 m sec⁻¹) psychrometer, dew point sensor or infrared analyzer	$\%$ RH or dew point or $g\ m^{-3}$	At top of plant canopy in center of plant growing area.	Once during each light and dark period, taken at least 1 hr after light changes. (Hourly measurements over the course of the study, advisable)	Average of once daily readings for both light and dark periods with range of diurnal variation over the period of the study (or average of hourly values if taken).
Air velocity:	$m\ s^{-1}$	At top of plant canopy. Obtain maximum and minimum readings over plant growing area.	At start and end of studies. Take 10 successive readings at each location and average.	Average and range of readings over containers at start and end of the study.
Carbon dioxide:	m moles m^{-3}	At top of plant canopy.	Hourly over the period of the study.	Average of hourly average readings and range of daily average readings over the period of the study.

Parameter	Units	Measurements			
		Where to take	When to take	What to report
Watering:	ml	---	At times of additions.	Frequency of watering. Amount of water added per day and/or range in soil moisture content between waterings.
Substrate:	---	---	---	Type of soil and amendments Components of soilless substrate.
Nutrition:	*Solid media:* $kg\ m^{-3}$ *Liquid culture:* micro **Nutrients**: $m\ moles\ l^{-1}$ macro **Nutrients**: $\mu\ moles\ l^{-1}$	---	At times of nutrient additions.	Nutrients added to solid media. Concentration of nutrients in liquid additions and solution culture. Amount and frequency of solution addition and renewal.
pH:	pH units	In liquid slurry for soil and in solution of liquid culture.	Start and end of studies in solid media. Daily in liquid culture and before each pH adjustment.	Mode and range during study.
Conductivity:	$dS\ m^{-1}$ *(decisiemens per meter)*[a]	In liquid slurry for soil and in solution of liquid culture.	Start and end of studies in solid media. Daily in liquid culture.	Average and range during study.

[a] $1dSm^{-1} = 1mho$

this USDA/SEA Committee has been toward development of guidelines
that will be as complete as possible and useful and acceptable to
investigators. It is recognized that scientists are likely to
accept only those recommendations that are not overly complex
and that can be followed with commercially available instrumentation at reasonable cost. Certain guidelines that more accurately
describe the plant environment have been outlined but have only
been encouraged at present because they involve excessive time or
cost. It is hoped that publication of these desired guidelines
will provide encouragement for the development of improved
instrumentation so that these proposed measurements can be readily
taken on a regular basis.

The committee has been encouraged to adopt SI units by
recommendations from scientists in many countries. Wide adoption
of SI units was considered to have great merit in reporting and
interpreting research results.

Members of the North Central Regional USDA/SEA Committee have
presented these guidelines and provided rationale for their
suggested form. Time was allowed in this conference for discussion, questions, and suggestions. The presentations on each
subject by the opening speaker and invited speakers were intended
to provide a common understanding of the significance of each
environmental parameter and the critical requirements in control
and sensing. The open discussions that followed the presentation
of the guidelines provided suggestions by the participants on
how these guidelines could be improved. These suggestions are of
particular value because they were contributed by individuals
representing a wide cross-section of manufacturers and users of
controlled environment facilities. We feel that participants in
this conference have collectively made a major contribution to
research on plant growth in controlled environments.

The effectiveness of this conference will be determined by how universally the suggested guidelines and successive modifications are accepted by plant scientists. The widespread acceptance of guidelines will require that they be formally accepted by various plant science societies and that journal editors be encouraged to require adherence to the guidelines in manuscripts submitted for publication.

REFERENCES

ASHS Committee on Growth Chamber Environments (1972). Guidelines for reporting studies conducted in growth chambers. *Hortscience 7*, 239.

ASHS Committee on Growth Chamber Environments (1977). Revised guidelines for reporting studies in controlled environment chambers. *Hortscience 12*, 309-310.

Chouard, P., and de Bilderling, N. (1972). II. Brief analysis of the proceedings of the Symposium: Use of phytotrons and controlled environments for research purposes. Durham-Raleigh (USA) 22-27 May, 1972. Phytotronics Newsletter No. 3 pp. 3-24. Phytotron - C.N.R.S. Gif-sur-Yvette, France.

Committee on Plant Irradiation of the "Nederlandse Stichting voor Verlichtingskunde" (1953). Specification of radiant flux and radiant flux density in irradiation of plant with artificial light. *J. Hort. Sci. 28*, 177-184.

Committee on Plant Irradiation of the "Nederlandse Stichting voor Verlichtingskunde" (1955). The determination of the irradiance in various spectral regions for plant irradiation practice. *J. Hort. Sci. 30*, 201-207.

Evans, L. T., ed. (1963). "Environmental Control of Plant Growth." Academic Press, New York.

Hudson, J. P., ed. (1957). "Control of the Plant Environment." Butterworth, London.

Kramer, P. J., Slatyer, R. O., and Hellmers, H. (1972). Phyto-
    trons and environmental physiology. *Nature and Resources*
    8(4), 13-16. Paris, UNESCO.

Rees, A. R., Cockshull, K. E., Hand, D. W., and Hurd, R. G. (1972).
    "Crop Processes in Controlled Environments." Academic Press,
    London.

White, F., ed. (1963). Proceedings of a symposium on engineering
    aspects of environmental control for plant growth. Common-
    wealth Scientific and Industrial Research Organization.
    Melbourne, Australia.

RADIATION

K. J. McCree

Department of Soil and Crop Sciences
Texas A & M University
College Station, Texas

INTRODUCTION

There have been many analyses of radiation measurements for plant growth studies in controlled environments (for example Bickford and Dunn, 1972; Downs and Hellmers, 1975; McCree, 1966, 1971, 1972a, 1972b, 1973; Norris, 1968). This paper will primarily review the basic principles of radiation measurement. Some useful terms and units will be defined. Photometric terms will not be discussed, because they are not relevant to plant growth. International Standard (SI) units will be used exclusively because they are the only ones that are recognized throughout the world (Incoll, Long, and Ashmore, 1977; Page and Vigoureux, 1972) and because they are logical and easily understood. The problems involved in applying basic physical principles to radiation measurements in plant growth studies, and in standardizing such measurements, will then be discussed.

BASIC PRINCIPLES OF RADIOMETRY

The basic quantity in radiometry is *energy,* the SI unit of which is the *joule* (J). Radiant energy is of most interest when

it is being transferred, in the form of an electromagnetic wave.
Then the relevant quantity is the *radiant power, in joules per
second (watts, W)*. Radiant flux leaves a source, is transmitted
through space, and is absorbed, reflected, or transmitted by a
receiving surface, which acts in turn as a second source, and so
ad infinitum. Fluxes from different sources are strictly
additive.

In the special case of an infinitely small source (a "point
source"), the flux emitted into unit solid angle of space (one
steradian) is called the *radiant intensity* (unit, W sr^{-1}), while
in the more general case of an extended source, the flux emitted
by *unit projected area* of surface into unit solid angle of space
is called the *radiance* (unit W m^{-2}sr^{-1}). The intensity and radiance always depend on the direction from which a source is viewed.

High intensity discharge lamps are examples of (near) point
sources, while fluorescent lamps are examples of surface sources.
The radiance of the surface source or the intensity of the point
source determines how much radiant flux (per unit solid angle, in
a given direction) can be obtained from the source. When a discharge lamp is used in a reflector, the radiance of the surface
of the reflector determines the available flux. Likewise, the
walls of growth chambers act as secondary sources, often with
multiple reflections between lamps and plants. Therefore the
properties of these reflecting surfaces are very important in
chamber design. Specular reflectors (mirrors) of high reflectance give the greatest uniformity.

The radiant flux incident on a surface from all directions,
per unit area of surface, is called the *irradiance* (unit, W m^{-2}),
while the analogous quantity for the flux leaving a surface is
called the *radiant exitance*. When a parallel beam of radiation
of given cross-sectional area spreads over a surface, the area
that it covers is inversely proportional to the cosine of the
angle between the beam and a plane normal to the surface. Therefore, the irradiance on the surface is proportional to the cosine

of this angle. If the response of a radiometer to radiation
coming from different directions does not follow the same law,
it will not measure true irradiance. It is said to have poor
"cosine correction".

All of the above quantities are also functions of the
wavelength of the radiation. The same terms are used when wave-
length-dependent properties are being specified, with the
addition of the prefix "spectral". For example, the *spectral
irradiance* at 400 nm is the irradiance per unit wavelength
interval, in a band centered on 400 nm (unit $W\ m^{-2}\ nm^{-1}$). Spectral dependence is implicit in all radiometric measurements,
since the wave band is never infinite.

Even though the wavelength interval specified in the spectral
irradiance unit is one nanometer, the actual wavelength interval
between data points can be greater than this. It is only
necessary that the spectral irradiance does not vary appreciably
over the stated interval. The size of the band of wavelengths
accepted by the measuring instrument (the "bandwidth") should
always be specified. This is particularly important when a spectroradiometer that has been calibrated with a continuum source
(incandescent lamp) is being used to measure the flux from a
monochromatic line source (Henderson and Hallstead, 1972), since
the response of a spectroradiometer is proportional to the band
width in the case of the continuum but not in the case of the
line. When line and continuum spectra appear together, as in the
case of a fluorescent lamp, the usual convention is to plot the
line as a rectangle, the area of which represents its spectral
irradiance when integrated over the same wavelength interval as
is being used for the continuum plot (Fig. 1a). This whole
problem would disappear if the rectangular plotting convention
were used for all spectroradiometric plots (Fig. 1b, 1c).

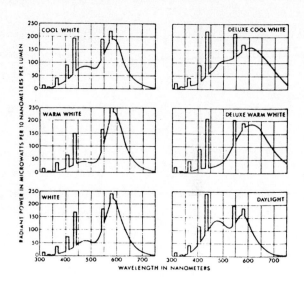

(a)

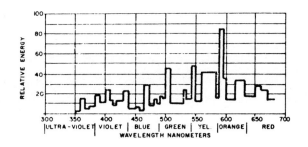

(b)

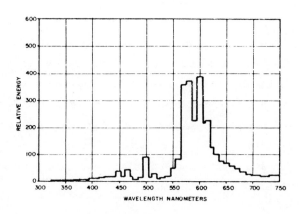

(c)

The most versatile and accurate spectroradiometers employ prisms or gratings to disperse the radiation. The bandwidth can be varied by varying the exit slit width. Wedge interference filters have been used in many portable instruments, while others use nondispersive colored or interference filters with a fixed wave band (Norris, 1968). Thus the bandwidth in a spectroradiometer will vary from one to a hundred nanometers, depending on the instrument. The optimum trade-off between bandwidth and cost is a topic that merits attention at this conference.

APPLICATION OF PRINCIPLES TO PLANT GROWTH STUDIES

This section will present some examples of problems that arise in attempting to specify the amount of radiation available to plants. These problems are not unique to controlled-environment studies, though they are exacerbated there, because there are so many different ways to control the environment. Perhaps investigators can solve some of their problems by adopting the

FIGURE 1. (a) Spectral distribution of the radiant power emitted by fluorescent lamps. The power emitted by the mercury line radiation per lumen is the area under the appropriate rectangular block, while the power emitted by the phosphor continuum radiation per lumen and per 10 mm wavelength interval is the height of the curve, at a given wavelength. (b) Relative spectral distribution of radiant power emitted by a phosphor-coated metal halide lamp. (c) Relative spectral distribution of radiant power emitted by a high-pressure sodium lamp. In Fig. 1(b) and (c) the band width is indicated by the width of the rectangles. (From Bickford and Dunn, 1972). (Original sources of data: Fig. 1(a) IES Lighting Handbook, Fig. 1(b) GTE Sylvania Inc., Fig. 1(c) General Electric Company).

techniques used by researchers in the field, where variability must be accepted and all measurements simplified. Spectral and temporal integration of data are normal practices in field studies.

Over the years, there has been a great deal of discussion about whether or not irradiance on a flat horizontal surface above the plants is the best measure of the density of radiant flux available to plants. The flux into a point (the so-called "spherical irradiance") has sometimes been considered superior, and in fact is being used in biological oceanography. The best solution probably is to retain the conventional irradiance measurement, while developing theoretical or empirical relationships between this irradiance and the flux absorbed by plants in a given geometrical situation (Monteith, 1975). The absorbed flux is, after all, the true parameter of interest in studies of plant growth. In field studies, the leaf area index (LAI) is the basis for such relationships (Table 1), but one seldom sees LAI values reported in controlled-environment studies. Why not? Are they all single-plant studies? Should they be?

TABLE 1. *Leaf Area Index (LAI) for 95% Absorption of Photosynthetically Active Radiation (LAI 95%) for Selected Crop Canopies. (From Monteith, 1969).*

CROP	LAI 95%
Clover (Trifolium repens L.)	2.9
Cotton (Gossypium hirsutum L.)	2.9
Beans (Phaseolus vulgaris L.)	3.8
Barley (Hordeum distichum L.)	4.7
Bullrush millet (Pennisetum typhoides L.)	5.5
Sorghum (Sorghum bicolor L. Moench)	6.9
Ryegrass (Lolium rigidum L.)	10.4

The spectral properties of radiation have, of course, occupied center stage for a very long time, probably because they are so very obvious. Anyone can see that the color of a sodium lamp is "unnatural" - therefore it is often considered to be incapable of growing "natural" plants. Problems with other properties of the radiation, such as low irradiance, or the improper daylength (or with unrelated factors such as the nutrient solution) have generally been overlooked.

This reviewer's particular prejudices about spectral properties have been aired several times (McCree, 1966, 1971, 1972a, 1972b, 1973). In summary they are: that certain parts of the spectrum should be identified with specific physiological responses, and that simplified measures of the quantity of radiation available to plants in these spectral regions should then be developed. It does not seem practicable to refine these measures to the point of close correspondence with the action spectrum for the plant response in question. These views are essentially the same as those expressed by the Wassink committee nearly 30 years ago (Committee on Plant Irradiation Ned. St. Verl., 1953).

It is important to limit the number of wave bands to be specified, because if this is not done, none will be used. It is unrealistic to expect complete spectroradiometric data for experiments that are not photobiological in nature. Even if such data were universally available, they could not be used to interpret the results of an experiment unless action spectra for various possible plant responses were also universally known.

Photosynthetically active radiation is by now well established as the first and probably most important plant-based radiometric quantity. The Committee on Crop Terminology of the Crop Science Society of America has developed a consistent set of definitions for use by scientists in the United States working in photosynthesis research (Shibles, 1976). These are:

Photosynthetically Active Radiation (PAR):
Radiation in the 400-700 nm waveband.
Photosynthetic Photon Flux Density (PPFD):
Photon flux density of PAR. The number of
photons (400 to 700 nm) incident per unit
time on a unit surface.
Photosynthetic Irradiance (PI): Radiant
energy flux density of PAR. The radiant
energy (400-700 nm) incident per unit time
on a unit surface.

The committee recommended units of $nE\ s^{-1}\ cm^{-2}$ and $nW\ cm^{-2}$
for the PPFD and PI respectively, but $\mu E\ s^{-1}\ m^{-2}$ and $W\ m^{-2}$ would
be more consistent with SI recommendations. Although the
Einstein (E) is not a defined SI unit, it is very well established in photobiology as a term for Avogadro's number (a mole)
of photons (Rabinowitch, 1951). Photochemical usage appears to
be inconsistent. In some texts the Einstein is said to be the
quantity of energy (in joules) carried by Avogadro's number of
photons (see Incoll, Long, and Ashmore, 1977 for references) while
in others the photobiological usage is followed (Arnold *et. al.*,
1974). On balance there seems little reason to abandon the use
of the Einstein in the PPFD unit, as was suggested by Incoll,
Long, and Ashmore, 1977).

Since the primary event in photosynthesis is a photochemical
one, it would be expected that photosynthetic responses of leaves
to radiation of different spectral qualities would follow PPFD
more closely than they would follow PI, and this is borne out by
the data (McCree, 1972b, Inada, 1976). The response of an
"average leaf" of a crop plant followed PPFD to within ±4%, but
PI to within only ±9%, for a representative range of artificial
light sources (Table 2). This would seem to be a good reason for
favoring use of PPFD over PI.

TABLE 2. *Calculated Photosynthetic Rate of an "Average Leaf",
per unit Flux of Incident PAR, when the Flux is measured in PI or
PPFD units (normalized to 1.00 for natural sun + sky radiation).
(From McCree, 1972b).*

Flux unit	Metalarc	Light Source CW Fluorescent	Lucalox	Incandescent
PI	1.02	0.97	1.14	1.15
PPFD	1.01	0.97	1.05	1.04

The second most discussed action of radiation on plants is
its effect on plant development. Several different photoreceptors
are involved, but phytochrome is best known. Phytochrome exists
in two interconvertible forms, the far-red absorbing form, P_{fr},
and the red absorbing form, P_r. The absorption spectra of the
two forms overlap at wavelengths below 700 nm, but the peaks lie
at 730 and 660 nm respectively (Fig. 2a). Hartmann (1966) calculated the fraction of the phytochrome that would be in the
morphogenically active form at equilibrium (P_{fr}/P_{total}) when a
pure solution is irradiated by monochromatic radiation of different wavelengths (Fig. 2b). The fraction is close to zero for
wavelengths above 700 nm, and varies between 0.5 and 0.8 for
wavelengths between 300 and 700 nm.

Unfortunately, it is not possible to measure the equilibrium
phytochrome fraction in green leaves. In dark grown plants
exposed to a range of natural and artificial light sources, the
fraction has been found to be close to that predicted from
Hartmann's curves and the spectral irradiance (Smith, 1975;
Morgan and Smith, 1978). Also, there was a close relationship
between P_{fr}/P_{total} and the ratio of the photon flux densities at
660 and 730 nm (PFD_{660}/PFD_{730}) (bandwidth 10 nm). Furthermore,
a good negative linear correlation between P_{fr}/P_{total} and the
logarithm of the rate of stem extension of the arable weed

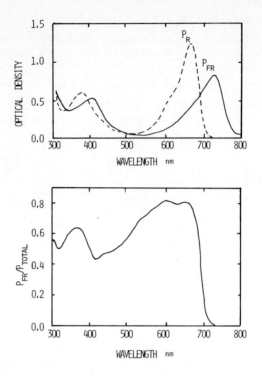

*FIGURE 2. (a) Absorption spectra of the red and far-red
absorbing forms of phytochrome (P_r and P_fr). (b) Proportion of
phytochrome expected to be in the far-red absorbing form at
equilibrium, when a solution is irradiated with monochromatic
radiation at different wavelengths. (After Hartmann, 1966).
(Reprinted with permission of Pergamon Press, Ltd.).*

Chenopodium album was demonstrated (Holmes and Smith, 1975, 1977
a, b; Morgan and Smith, 1978). The most striking variations in
the photon flux ratios and in phytochrome equilibria in etiolated
tissue samples were observed within plant canopies, as would be
expected from the fact that leaves absorb very strongly at 660 nm
but very weakly at 730 nm (Monteith, 1976). It was concluded
that phytochrome was the photoreceptor that enabled the weed to
respond appropriately to shading by other plants.

The actual phytochrome equilibria in green tissues would be
expected to be different from those observed in the experiments.
Because of the strong preponderance of wavelengths less than

700 nm in the spectra of lamps used in controlled environment
chambers, the phytochrome equilibrium will be near the maximum
P_{fr}/P_{total} of 0.8. Some increase in the rate of stem extension
of *Chenopodium album* resulted from adding incandescent to
fluorescent lamps, but a special far-red source was needed to
produce a greater range of values of P_{fr}/P_{total} for experimental
purposes (Morgan and Smith, 1976, 1978) (Fig. 3). The experiments
were performed at low PPFD (100 $\mu E\ m^{-2}\ s^{-1}$).

The results of these experiments by Smith and his colleagues
are very encouraging. Although the relationships that they dis-
covered are empirical, they are based on expectations that are
reasonable from a physiological point of view. Perhaps it may
soon be possible to proceed with a specification for "photomorphogenically active radiation" based on the ratio of two
photon flux densities.

Of course it goes without saying that all photomorphogenesis
experiments should be done at constant *photosynthetic* photon flux
densities, since photosynthesis is the primary process that
determines growth. If possible, the *absorbed* fluxes rather than
the *incident* fluxes should be equalized. The PPFD should be
comparable with that found outdoors.

Ultraviolet radiation cannot be entirely neglected, since it
is potentially very damaging to plants, especially at wavelengths
less than 300 nm. For obvious reasons there is very little such
radiation in the spectra of lamps sold for general lighting
purposes. Furthermore, such radiation is strongly absorbed by
glass and most plastics used for barriers. Hence the irradiances
of ultraviolet radiation in most controlled environment chambers
are likely to be negligible. When special sources such as xenon
lamps are used, the normal precautions that protect people should
suffice to protect the plants.

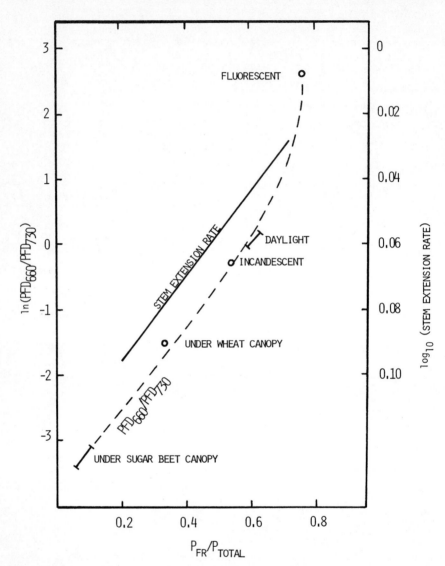

FIGURE 3. *Relationship between equilibrium active phytochrome fraction in dark-grown tissue (P_{fr}/P_{total}) and (a) the natural logarithm of the photon flux densities at 660 and 730 nm (broken line, left scale), for various natural and artificial light sources, and (b) the logarithm of the stem extension rate of Chenopodium album (solid line, right scale) when irradiated by various mixtures of fluorescent, incandescent and far-red light sources. (Data of Holmes and Smith, 1975, and Morgan and Smith, 1976, replotted by the author).*

STANDARDIZATION OF MEASUREMENTS

Discussion of specific instruments has been deliberately avoided up to this point because it is first necessary to decide *what* must be measured, and with what precision and what accuracy. Then it will be up to manufacturers to design instruments that meet the agreed specifications. The sad history of the continued use of photometric instruments in plant growth studies emphasizes the danger of taking the reverse approach, that is, basing specifications on available instruments.

Of course, it must be possible to meet specifications at reasonable cost, and, in order to do this, there must be input from specialists in radiometric instrumentation. There is surely no lack of such expertise in industry, but I wonder if it has been drawn on sufficiently in the past. There is no place for the novice in radiometry.

Since the purpose of using controlled environment facilities presumably is to be able to mimic outdoor condition, but in a more controlled fashion, it would seem logical to use the same instruments indoors that are used outdoors. Thus the *shortwave irradiance* (300 - 3,000 nm) would be measured with the conventional pyranometer, placed in a horizontal position above the plants. These instruments use thermal detectors that are inherently non-selective. Instruments based on photon detectors such as silicon cells are inherently selective and should not be used for this purpose.

Broad-band PAR measurements within the shortwave band are complicated by the existence of two rival measures, the photosynthetic irradiance (PI) and the photosynthetic photon flux density (PPFD). As noted above, PPFD is to be preferred. We do need more demonstrations of the superiority of PPFD in practice. Whichever unit is used, it is important to retain the prefix "photosynthetic" as an indicator of the wave band (400-700 nm in the United States). The Crop Science Society committee's nomenclature should be followed.

Since PAR sensors are much smaller than pyranometers, it becomes possible to determine the geometric properties of the radiation in more detail when PAR sensors are used. Again, the experience of field users can be used as a guide. Use of these sensors within plant canopies provides the most critical test of their ability to conform to the 700 nm cutoff point, since the spectral irradiance increases abruptly at 700 nm under a leaf (Monteith, 1976).

It probably is premature to base a firm specification for photomorphogenically active radiation on the available evidence on correlations between stem elongation rate and phytochrome equilibrium, and between phytochrome equilibrium and photon flux ratio (660/730 nm). However, it is clear that a ratio is more logical than an absolute quantity of far-red radiation, since two forms of phytochrome are involved. What would be an appropriate bandwidth for these measurements?

The *precision* of radiometric measurements (that is, the repeatability of measurements with any instrument) can be comparable with that of most other physical measurements (say ± 1%), but the *accuracy* (that is, capability to measure in known absolute units) is much less. Only the best-equipped standardizing laboratories can expect to achieve an accuracy of ± 5% or better. It is suggested that ± 10% would be a good value to air at for careful but routine measurements made in conventional facilities. Thus, when several different makes of PPFD meters are placed beside each other and exposed to different sources of radiation, their readings should agree within ± 10%. Even then, the test would have to be limited to broad-spectrum "white" light sources and diffuse radiation. The use of narrow-band sources or highly directional beams would probably result in much greater differences. Intercomparisons of different types of meters (e.g. pyranometers versus PI meters, PI meters versus PPFD meters) are not recommended. The results are very specific to the light source and are difficult to interpret. It is better to accept

the fact that these meters measure inherently different properties of the radiation.

It is unrealistic to expect the average user of a radiometric
instrument to go much beyond a simple comparison of similar types
of meters. The best that can be done is to familiarize him with
the pitfalls of such a comparison. A large controlled-environment
facility will no doubt employ a technician whose job it is to
repair and calibrate instruments, and such a technician could
handle basic radiometric calibrations, given the right equipment
and a course in radiometry at a recognized training center. One
solution to the chronic problems in this area might be for such
a central facility to accept instruments from other users for
regular periodic adjustment and calibration by a technician
trained in radiometric measurements.

REFERENCES

Arnold, D. R., Baird, N. C., Bolton, J. R., Brand, J. C. D.,
    Jacobs, P. W. M., de Mayo, P., and Ware, W. R. (1974).
    "Photochemistry: An Introduction." Academic Press, New York.
Bickford, E. D., and Dunn, S. (1972). "Lighting for Plant Growth."
    Kent State University Press, Kent, Ohio.
Committee on Plant Irradiation Ned. St. Verl. (1953). Speci-
    fication of radiant flux and radiant flux density in
    irradiation of plants with artificial light. *J. Hort. Sci.*
    28, 177-184.
Downs, R. J., and H. Hellmers. (1975). "Environment and the
    Experimental Control of Plant Growth." Academic Press, New
    York.
Hartmann, K. J. (1966). A general hypothesis to interpret high
    energy phenomena on the basis of phytochrome. *Photochem.*
    Photobiol. *5,* 349-366.
Henderson, S. T., and Hallstead, M. B. (1952). The spectro-
    photometry of light sources. *Brit. J. App. Phys. 3,* 255-259.

Holmes, M. G., and Smith, H. (1975). The function of phytochrome
    in plants growing in the natural environment. *Nature 254,*
    512-514.
Holmes, M. G., and Smith, H. (1977a). The function of phytochrome
    in the natural environment. III. Measurement and calculation
    of phytochrome photoequilibria. *Photochem. Photobiol. 25,*
    547-550.
Holmes, M. G., and Smith, H. (1977b). The function of phytochrome
    in the natural environment. IV. Light quality and plant
    development. *Photochem. Photobiol. 25,* 551-557.
Inada, K. (1976). Action spectra for photosynthesis in higher
    plants. *Plant Cell Physiol. 17,* 355-365.
Incoll, L. D., Long, S. P., and Ashmore, M. R. (1977). SI units
    in publications in plant science. *Curr. Adv. Plant Sci. 9,*
    331-343.
McCree, K. J. (1966). A solarimeter for measuring photo-
    synthetically active radiation. *Agric. Meteorol. 3,* 353-366.
McCree, K. J. (1971). Significance of enhancement for calculations
    based on the action spectrum for photosynthesis. *Plant
    Physiol. 49,* 704-706.
McCree, K. J. (1972a). The action spectrum, absorptance and
    quantum yield of photosynthesis in crop plants. *Agric.
    Meteorol. 9,* 191-216.
McCree, K. J. (1972b). Test of current definitions of photosyn-
    thetically active radiation against leaf photosynthesis data.
    Agric. Meteorol. 10, 443-453.
McCree, K. J. (1973). A rational approach to light measurements
    in plant ecology. *Curr. Adv. Plant Sci. 3,* 39-43.
Monteith, J. L. (1969). Light interception and radiative exchange
    in crop stands. *In* "Physiological Aspects of Crop Yield"
    (J. D. Eastin, ed.), pp. 89-111. American Society of Agronomy,
    Madison, Wisconsin.
Monteith, J. L. (1975). "Principles of Environmental Physics."
    Edward Arnold, London.

Monteith, J. L. (1976). Spectral distribution of light in leaves
and foliage. *In* "Light and Plant Development" (H. Smith, ed.),
pp. 447-460. Butterworth, London.

Morgan, D. C., and Smith, H. (1976). Linear relationship between
phytochrome photoequilibrium and growth in plants under
simulated natural radiation. *Nature 262,* 210-212.

Morgan, D. C., and Smith, H. (1978). The relationship between
phytochrome photoequilibrium and development in light grown
Chenopodium album L. *Planta 142,* 187-193.

Norris, K. H. (1968). Evaluation of visible radiation for plant
growth. *Annu. Rev. Plant Physiol. 19,* 490-499.

Page, C. H., and Vigoureux, P. (1972). The International System
of Units (SI). National Bureau Standards Special Publication
330.

Rabinowitch, E. E. (1951). "Photosynthesis and Related Processes."
Interscience, New York.

Shibles, R. (1976). Committee Report: Terminology pertaining to
photosynthesis. *Crop Sci. 16,* 437-439.

Smith, H. (1975). "Phytochrome and Photomorphogenesis" , p. 151.
McGraw Hill, London.

RADIATION: CRITIQUE I

Robert J. Downs

Phytotron
North Carolina State University
Raleigh, North Carolina

One of the objectives of a discussant is to disagree as often
as possible with the principal speaker in order to stimulate
animated discussion from the floor. Since McCree discussed
physical principles and their application to plant growth,
little disagreement is practicable. It is possible however to
expand upon some of McCree's comments, especially from the point
of view of plant growth and development.

Lowry (1969) reminds us that no matter how we choose to
describe the environment, whether by direct or indirect measurements, we must keep in mind the objective of the task. The
usual objective of plant research in controlled environment
chambers is to describe responses to the environment and to
explain and understand how the responses occur. Even when plant
growth chambers are used to provide constant conditions for
growing standard plants, the objective is still monitoring the
plant response to experimental treatments. In no case is
describing the environment the objective; it is only the means of
reaching it. Therefore investigators must guard against
becoming so engrossed in the physical measurement of radiant
energy that they obscure the biological purpose of using it.

ILLUMINANCE

Undeniably the measurement of illuminance fails to properly
describe radiant energy as it is used by plants. A number of
authors, including Bickford and Dunn (1972) and McCree (1973)
have explained in detail that this failure results from the
spectral sensitivity of the illumination meter being attuned,
quite deliberately, to human vision. Therefore, all regions of
the spectrum utilized by plants are not included in the measurement; and those that do fall within the measured spectral band
are not detected equally well. Biologists have always been aware
of the undesirability of illumination measurements, but have been
directed to their use by economics and/or the need to make
their research results applicable to commercial growers.

TOTAL ENERGY

Many investigators who frown on the use of luminous flux
density are advocates of total energy measurements. The reason
for this is not clear but it seems to be based on obscure
reasoning that a gram-calorie is more meaningful than a foot
candle; or as McCree suggested, that it is logical to use the
same set of instruments indoors as is used outdoors. Such logic,
of course, is valid only if the instruments are being used for
the same purpose. Generally outdoor instruments are used in
meteorology, and meteorological data are used for studying
weather. Consequently a pyrheliometer or pyranometer that
provides satisfactory radiation data for solar constant work or
for meteorology may be completely unsatisfactory for plant
physiology.

It seems rather obvious that total energy, even over the
relatively small range of 300 to 3000 nm, presents a completely
unrealistic picture of plant growth lighting. For example,
recent measurements made in a fluorescent + incandescent

lamped controlled-environment room show that the 2400 W of
incandescent lamps produce about the same total irradiance
as the 6020 W of fluorescent lamps (Table 1). Can we conclude
therefore that since total energy is the same under the two
light sources, plant growth and development will also be equal?
Obviously the answer is no. Illuminance ratios of 1:12 represent
the biological effectiveness of the light sources much more satis-
factorily than the 1:1 ratio of total irradiance.

TABLE 1. *Change in Radiant Energy Factors due to the Addition
of Incandescent Lamps in a Plant Growth Chamber.*

Fluorescent	Incandescent	Increase due to incandescent lamps
6020 watts	2400 watts	40% watts
470 hlx	37 hlx	8% illuminance
130 $W\ m^{-2}$	130 $W\ m^{-2}$	100% total energy (400-1500 nm)
650 $\mu E\ m^{-2} s^{-1}$	70 $\mu E\ m^{-2} s^{-1}$	11% PAR (400-700 nm)
1 $W\ m^{-2}$	9 $W\ m^{-2}$	900% PR (700-850 nm)

THERMAL RADIATION

   As a measure of the energy absorbed and used by plants,
inclusion of long wavelengths is likely to encompass much energy
that is not used by the plant photoreactions. This does not mean
that total energy measurements are not useful; although their
value would seem to be limited without additional data that
would define the total energy budget and its components
(Reyenga and Dunin, 1975). Certainly thermal radiation does
influence plant growth and may, directly or indirectly, influence
the effectiveness with which photosynthetic energy is used.
However, this does not provide sufficient cause to combine
thermal and photosynthetic energy in a single measurement. On

the other hand there seems to be considerable merit in measuring
thermal radiation separately from visible and near visible
radiant energy. For example, Biamonte (Table 2) found that very
high photon flux densities with large amounts of thermal radiation may not only fail to increase growth over that attained with
more moderate energy levels, but can actually reduce growth and
under some conditions be lethal. This is an effect of radiant
heat that does not necessarily result from raising substrate,
plant or air temperatures to the thermal death point. Thus
Biamonte's (1972) geranium seedlings, which were killed by the
radiant heat that accompanied the high photon flux densities,
survived raising the leaf temperatures to the same levels without
the radiant heat. It is not surprising therefore to note that
investigators such as Warrington (1977), who have had considerable
experience with high intensity discharge lamps which produce
high levels of radiant heat, consider a thermal filter imperative
when such sources are used for plant growth chamber lighting.

TABLE 2. *Effect of Radiant Flux Density on the Time of Inflorescence Development and Heights attained by Carefree White
geraniums. (From Biamonte, 1972).*

Radiant flux density $\mu E\ m^{-2}\ s^{-1}$	Stage 1[a]		Stage 2		Stage 3	
	Days	Height (cm)	Days	Height (cm)	Days	Inflores-cence Number
370	94	40	121	50	133	2.6
740	40	13	63	25	79	3.8
1480 + H_2O	46	16	70	27	89	4.5
1480	Plants dead-----------------------------					

[a]*Soil temperature with and without water bath 27 and 33° C;
air temperature with and without water bath 27 and 31° C;
leaf temperature with and without water bath 25 and 29° C.*

PHOTOSYNTHETICALLY ACTIVE RADIATION

As McCree already noted, photosynthetically active radiation
(PAR) is rapidly becoming the accepted radiometric quantity for
use in plant growth lighting. Some efforts have been made to
establish 400-700 nm as the standard wave band for PAR (Shibles,
1976). However, neither the 400-700 nm limits or PAR itself has
been universally accepted. In the USSR, for example, PAR seems
to be advocated but over a 380 to 710 wave band (Khazanov, 1978).
Elgersma and Meijer (1976) have gone somewhat further and propose
a phytolumen per mW. Since the average photosynthesis action
spectrum shows a lower effectiveness of blue quanta, the
efficiency of an irradiance of a given spectral distribution for
plant growth would depend on the spectral energy distribution of
the light source. Thus the phytolumen per mW is a conversion
factor used as a unit of plant growth. Although this approach
may be an unnecessary refinement in practice, Elgersma and
Meijer (1976) suggest that it disposes of two highly debatable
criteria for plant growth lighting: (1) That a good lamp for
plant growth must have a spectral distribution similar to that of
the sun. The phytolumen method shows this is not true since the
conversion factor for plant growth is actually worse in sunlight,
0.67 phytolumen per mW than it is for many artificial sources
currently in use, such as fluorescent lamps with a conversion
factor of 0.79 phytolumen per mW, (2) That the spectral energy
distribution of the light source should have the same form as
the plant sensitivity curve. This is not necessarily so either
because, while a good conversion factor of 0.73 phytolumen per
mW is produced, it is still less than that for a number of
existing lamps.

PHOTOMORPHOGENESIS

Measurements of photon flux density between 400 and 700 nm,
while providing useful information about potential photosynthesis,
do not indicate the kind of growth to expect under different kinds
of light sources. For example, plant growth under fluorescent or
incandescent lamps equilibrated to provide equal photon flux
densities of PAR result in significantly different amounts of
plant growth and chlorophyll (Table 3). Considering all the
facts available it seems strange to find opinions to the effect
that white light from fluorescent lamps in growth chambers
provides all the radiation necessary for the reactions that
determine such plant responses as germination, flowering, and
presumably growth.

It is a well established fact that addition of some light
from incandescent lamps to the main light source of a plant
growth chamber, whether arc lamps (Parker and Borthwick, 1949);
fluorescent lamps (Went, 1957); (Cathey, Campbell, and Thimijan,
1978); or metal halide lamps (Warrington, Mitchell, and
Halligan, 1976), results in significant increases in plant dry
weight, starch, and sugars (Table 4). More recent data (Table 5)
confirm these earlier results and show clearly that the incandescent supplement to the fluorescent light increases plant
fresh weight, stem length, and rate of flowering.

McCree briefly discussed phytochrome and its importance in
plant development. Nevertheless others seem to harbor the idea
that there are few data to demonstrate either a requirement for
a balance or the necessity of a period of only red or far red
radiation at the initiation or conclusion of the light period.
The importance of far red was first noticed by C. W. Doxator
(Borthwick and Parker, 1952). For several years Doxator had
successfully grown sugar beets for seed during the winter using
incandescent-filament lamps to provide the long days required

TABLE 3. *Developmental Changes in Tripleurospermum maritimum and Chenopodium album over 15 days at 25°C under Fluorescent or Incandescent Light Sources.* (From Holmes and Smith, 1975)

Property	T. maritimum		C. album	
	Fluorescent[a]	Incandescent[a]	Fluorescent[a]	Incandescent[a]
Height (cm)	29.7	59.4	15.0	28.4
Internode length (cm)	0.8	3.5	-----	-----
Leaf dry weight (g)	0.48	0.46	0.34	0.31
Stem dry weight (g)	0.33	0.58	0.10	0.20
Leaf area (dm^2)	-----	-----	1.07	0.78
Chlorophyll a+b (mg/g FW)	-----	-----	112.3	93.8

[a] Light sources adjusted to produce equal quantum flux densities in the 400-700 nm wave band P_{FR}/P_{TOT} fluorescent 0.71, incandescent 0.38.

TABLE 4. *Yield of Biloxi Soybeans after 4 Weeks Growth under Carbon Arc Lamps with and without Incandescent Lamps. (From Parker and Borthwick, 1949).*

Property per plant	Arc	Arc + incandescent
Dry weight (g)	1.62	2.45
Starch (mg)	36	94
Sugars (mg)	61	114

TABLE 5. *Growth of Lettuce, Bean, and Corn Plants under Fluorescent Lamps with and without Incandescent. (From Rajan, Betteridge, and Blackman, 1971; Deutch and Rasmussen, 1974).*

Light source	Bolting index[a] Lettuce	Fresh weight (g) Pinto bean	Corn	Stem length (cm) Pinto bean
Fluorescent	9.8	3.5	3.6	12.9
Fluorescent plus incandescent	19.9	7.0	6.2	20.4

[a]*Bolting index is cm stem height.*

for the flowering of these plants. He then built a new greenhouse complete with the most efficient lamp of the time, the
fluorescent one, for photoperiod control. This seemed logical
because action spectra had shown red radiation to be the most
efficient portion of the spectrum for photoperiod control and the
red energy produced from the fluorescent lamps would be greater
than that obtained from incandescent lamps. Nevertheless the
sugar beets failed to flower under the fluorescent light.
Flowering and successful seed production could only be attained
by reverting to the use of incandescent lamps. Controlled
experiments subsequently confirmed Doxator's experience (Table 6).

TABLE 6. *Effect of Light Source on Flowering of Sugar Beets.
(From Borthwick and Parker, 1952).*

Supplemental light source	Sugar beets with flower stalks per lot of 12
Incandescent	11
Fluorescent	
Warm white	0
Soft white	0
Cool white	0
Daylight	0
Agricultural[a]	0
None	0

[a]*The agricultural lamps provided much more red than the white
lamps. These original plant growth lamps were similar to Grolux
lamps.*

Later studies showed clearly that the kind of light source used for photoperiod control altered the amount of flowering stimulus provided by a given daylength (Table 7); an effect that seems at odds with action spectra for flowering induction. The kind of light used for photoperiod control also regulates the amount of growth attained by woody plants as a function of day-length (Table 8). Similar plant responses, including elongation of internodes of herbaceous plants (Table 9), are obtained in plant growth chambers equipped with more than one kind of light source when one light source is turned off before the other.

Detailed studies showed that these plant responses to different kinds of light sources resulted from the amount of far red emitted by the lamp. As early as 1956 it was shown that the degree of elongation of internodes is a function of the relative amount of phytochrome in the P_{FR}, far red absorbing form (Fig. 1). Thus a few minutes of far red radiation at the close of the light period removes most of the P_{FR} which alters growth (Table 10) and drastically increases internode length (Fig. 2). The ratio of red and far red absorbing forms of phytochrome also has been shown to regulate cold acclimation (McKenzie, Weiser, and Burke, 1974) and synthesis of alkaloids and phenolic compounds (Tso, Kasperbauer, and Sorokin, 1970).

McCree pointed out that since the spectrum of lamps used in controlled-environment rooms is predominantly in wavelengths less than 700 nm, the P_{FR}/P_{TOT} equilibrium may be near 0.8. How-ever half maximum physiological change has been obtained by shif-ting this equilibrium ratio by as little as 0.1. Also, as the data presented in Table 5 show, the small change in the P_{FR}/P_{TOT} ratio due to the addition of incandescent lamps results in a marked biological response. It is suggested therefore that con-siderable evidence exists to support the statement that a des-cription of the radiant energy in controlled-environment rooms must include photomorphogenic as well as photosynthetic photon

TABLE 7. Effect of the Kind of Supplemental Light on Flowering Behavior of Little Club Wheat and Colsess Barley.

Supplemental light	Period to heading (days)		Stem length (cm)		Spike length (mm)		Spike weight (g)		Fertile grains per spike (no)	
	Wheat	Barley	Wheat	Barley	Wheat	Barley	Wheat	Barley	Wheat	Barley
None	a		----		----		----		----	
Fluorescent	91	98	85	57	56	45	0.44	0.22	27	5
Incandescent	70	54	91	62	40	40	0.84	0.88	25	13

[a] Spike beginning to appear after 160 days.

TABLE 8. Mean Increase in Growth of Pinus taeda Seedlings During 13 Weeks of 16 hr Photoperiods Consisting of 8 hr of Sunlight and 8 hr of 300 Lux Illumination from Fluorescent or Incandescent-filament lamps. (From Downs, 1957)

Kind of supplemental light	Stem length (mm)	Needle length (mm)
None (8-hour day)	41	0
Incandescent-filament	120	44
Fluorescent	58	24

TABLE 9. Effect of Light Quality at the Close of the Light Period on Growth of Tobacco Seedlings. (From Downs, 1975)

Variety	Light source[a]	Stem length (cm)	Fresh weight of leaves (g)	Stem Diameter (cm)
Coker 319	Fluorescent	13.0	25.0	4.1
	Incandescent	21.0	30.3	9.7
Coker 254	Fluorescent	13.0	30.6	5.2
	Incandescent	21.3	33.4	10.3
Coker 298	Fluorescent	10.6	27.0	3.6
	Incandescent	15.2	30.0	7.4
NC-2326	Fluorescent	10.9	28.2	4.6
	Incandescent	19.2	29.1	9.6
NC-98	Fluorescent	9.3	27.0	3.6
	Incandescent	16.8	24.6	8.9

[a]For 0.5 hr at the end of 8.5 hr of fluorescent plus incandescent at a photon flux density of PAR of 670 $\mu E\ m^{-2}\ s^{-1}$.

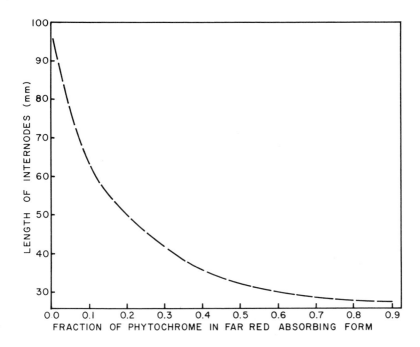

Figure 1. Internode lengths of Pinto beans as a function of phytochrome conversion. (From Hendricks, Borthwick, and Downs, 1956).

flux densities. Admittedly, ideal instrumentation is not available to more than a few investigators. In order to develop systems available to many, perhaps a measure of phytochrome action should be based, as McCree suggests, on an amalgam of detailed action spectra. Or perhaps photomorphogenic radiation could be measured by developing transducers to match the absorption spectra of the two forms of phytochrome. Or it may be equally significant to simply measure the incident energy in 20 nm bands at 660 and 735 nm. Whichever method is ultimately used, investigators can hardly afford to await its development. They can proceed at once to get an indication of the morphogenic potential of light sources by using existing, readily available, transducers that measure irradiance between 700 and 850 nm.

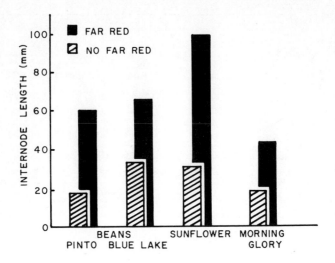

Figure 2. Effect of far red irradiation at the beginning of the dark period on internode elongation of bean, sunflower and morning glory plants. (From Downs, Hendricks, and Borthwick).

TABLE 10. Effect of Far Red (FR) at the End of the Light Period on Growth of Radish. (From Proctor, 1973)

	5 min FR[a]	No FR
Plant height (cm)	16.9	9.8
Leaf length (cm)	16.8	11.2
Leaf width (cm)	5.4	4.7
Fresh weight (g)	5.2	4.2
root	8.7	7.4 ns
top	5.2	4.2
Chlorophyll absorbance at 600 nm	0.177	0.344

[a] $0.87 \ \mu W \ cm^{-2} \ nm^{-1}$ at 750 nm.

REFERENCES

Biamonte, R. L. (1972). The effects of light intensity on the
initiation and development of flower primordia and growth
of geraniums. M.S. Thesis, Horticulture Department, North
Carolina State University, Raleigh, North Carolina.

Bickford, E., and Dunn, S. (1972). "Lighting for Plant Growth."
Kent State University Press, Kent, Ohio.

Borthwick, H. A., and Parker, M. W. (1952). Light in relation to
flowering and vegetative development. Rept. 13th Int. Hort.
Congr., London.

Cathey, H. M., Campbell, L. E., and Thimijan, R. W. (1978).
Comparative development of 11 plants grown under various
fluorescent lamps and different durations of irradiation
with and without additional incandescent lighting. J. Amer.
Soc. Hort. Sci. 103, 781-791.

Deutch, B. and Rasmussen, O. (1974). Growth chamber illumination
and photomorphogenic efficacy. I. Physiological action of
infra red radiation beyond 750 nm. Physiol. Plant 30, 64-71.

Downs, R. J. (1957). Photoperiodic control of growth and dormancy
in woody plants. In "The Physiology of Forest Trees"
(K. V. Thimann, ed.), pp. 259-538. Ronald Press, New York.

Downs, R. J. (1975). "Controlled Environments for Plant Research."
Columbia University Press, New York.

Downs, R. J., Hendricks, S. B., and Borthwick, H. A. (1957).
Photoreversible control of elongation of Pinto beans and
other plants under normal conditions of growth. Bot. Gaz.
118, 119-208.

Elgersma, O., and Meijer, G. (1976). Plant culture under artificial light. Proc. 4th European Orchid Congr., pp. 22-30,
Amsterdam.

Hendricks, S. B., Borthwick, H. A., and Downs, R. J. (1956).
Pigment conversion in the formative responses of plants to
radiation. Proc. Natl. Acad. Sci. 42, 19-26.

Holmes, M. G., and Smith, H. (1975). The function of phytochrome in plants growing in the natural environment. *Nature 254*, 512-514.

Khazanov, V. S. (1978). Photometers for phytotrons. *Phytotronic Newsletter 17*, 47-50.

Lowry, W. P. (1969). "Weather and Life." Academic Press, New York.

McCree, K, J. (1973). A rational approach to light measurements in plant ecology. *Curr. Adv. Plant Sci. 5*, 39-43.

McKenzie, J. S., Weiser, C. J., and Burke, M. J. (1974). Effects of red and far red light on the initiation of cold acclimation in *Cornus stolonifera*. *Plant Physiol. 53*, 783-789.

Parker, M. W. and Borthwick, H. A. (1949). Growth and composition of Biloxi soybean grown in a controlled environment with radiation from different carbon arc sources. *Plant Physiol. 24*, 345-358.

Proctor, J. R. A. (1973). Developmental changes in radish caused by brief end of day exposure to far red radiation. *Can. J. Bot. 51*, 1075-1077.

Rajan, A. K., Betteridge, B., and Blackman, G. E. (1971). Interrelationships between the nature of the light source, ambient air temperature and the vegetative growth of different species within growth cabinets. *Ann. Bot. 35*, 323-342.

Reyenga, W., and Dunin, F. X. (1975). An assembly for measuring components of the radiation balance. *Aust. CSIRO Div. Plant Ind. Field Stn. Rec. 14*, 17-23.

Shibles, R. (1976). Terminology pertaining to photosynthesis. *Crop Sci. 16*, 437-439.

Tso, T. C., Kasperbauer, M. J., and Sorokin, T. P. (1970). Effect of photoperiod and end-of-daylight quality on alkaloids and phenolic compounds of tobacco. *Plant Physiol. 45*, 330-333.

Warrington, I. J. (1977).   Lighting systems in controlled-environ-
    ment chambers.   *Proc. Workshop on Controlled Environment
    Cabinets,* Tech. Rept. 6, Plant Physiol. Div.   Palmerston
    North, New Zealand.
Warrington, I. J., Mitchell, K. J., and Halligan, C. (1976).
    Comparison of plant growth under four different lamp combina-
    tions and various temperature and irradiance levels.
    Agric. Meteor. 16, 231-245.
Went, F. W. (1957). "Environmental Control of Plant Growth."
    Chronica Botanica, Waltham, Massachusetts.

irradiance systems, the spectral irradiance and spectral
irradiance measurements, then he or she is ready to face the many
disappointments to be met in the market place where the cost,
equipment, and technology needed to fit these definitions do
not fit the project budget. Does the investigator quit or
compromise? The choice is his or hers, and it is frequently
the latter.

Bickford and Dunn (1972) emphasized the need for moving from
photometric terminology and measurement, to radiation terminology
and measurement in plant research. While photometric terms are
still being used, most measurements are now being expressed in
absolute terms (ASHS, 1977) and in the guidelines distributed at
this meeting. Some very useful suggestions have been made here
and elsewhere (Downs, 1975).

McCree has made some constructive suggestions for various
measurements and some of their practicalities. These suggestions
deserve attention and testing. He is to be commended for his
efforts and contributions at this conference and for his
pioneering work in plant radiation measurements. He was refining
the art and science of plant irradiation measurements (McCree,
1966) at a time when many investigators were still using foot
candles, or the SI photometric unit, lux (lumens per square meter).

As McCree stated, the SI unit for irradiance is $W\,m^{-2}$. There
may be the belief that this is the only legitimate SI unit that
can be used for irradiance. While it is a satisfactory unit for
high irradiance levels, it is a cumbersome one for expressing low
levels of irradiance which may be used in photomorphogenesis or
narrow spectral band width studies. In these latter studies the
smaller irradiance units, $mW\,cm^{-2}$ or $\mu W\,cm^{-2}$, are more convenient. To those who may believe that these are not SI units, it
pleases me to announce that they are because they contain the
officially accepted SI prefixes shown below.

For use with SI units there is a set of 16 prefixes (see
Table 1) to form multiples and submultiples of these units. It

RADIATION: CRITIQUE II

E. D. Bickford

Duro-Test Corporation
North Bergen, New Jersey

The major emphasis on radiation in this conference implies that it is an important parameter in plant research in controlled environments. It also implies that particular radiation sources are to be used and that radiation measurements are to be made.

In choosing a radiation source or instruments for measuring radiation the investigator must recognize the great difference between what is ideal and what is practical. For example, if the objective is to simulate the spectral irradiance of natural light (300-700 nm), the choice might be between sunlight-simulating fluorescent lamps and filtered high pressure xenon lamps. The cost and degree of sophistication of control gear are orders of magnitude apart. By the same token, there are great differences between very accurate spectroradiometers and PAR, PI, or PPFD meters. For some research problems a combination of the simplest sources and measuring devices is adequate; in other cases the most accurate and complex measuring devices may be inadequate for one reason or another.

If an investigator is considering conducting research in which radiation is an important parameter and which requires the use of radiation sources and making radiation measurements, he or she should think very seriously about methodology. When an investigator can resolve in his or her mind the explicit definitions of the

47

is important to note that the kilogram is the only SI unit with a prefix. Because double prefixes are not to be used, the prefixes in Table 1, in the case of mass, are to be used with gram (symbol g) and not with kilogram (symbol kg).

TABLE I.   SI Prefixes and Symbols *

Factor	Prefix	Symbol
10^{18}	exa	E
10^{15}	peta	P
10^{12}	tera	T
10^{9}	giga	G
10^{6}	mega	M
10^{3}	kilo	k
10^{2}	hecto	h
10^{1}	deka	da
10^{-1}	deci	d
10^{-2}	centi	c
10^{-3}	milli	m
10^{-6}	micro	μ
10^{-9}	nano	n
10^{-12}	pico	p
10^{-15}	femto	f
10^{-18}	atto	a

*(FR Doc. 77-31094 Filed 10-25-77; 8:45 a.m.)

For more information regarding the International System of Units, the investigator should contact the Office of Technical Publications, National Bureau of Standards, U.S. Department of Commerce, Washington, D. C. 20234, Jordan J. Baruch, Assistant Secretary for Science and Technology.

Most investigators are familiar and comfortable with controlled environments in which the radiation sources are some combination of cool white fluorescent and incandescent lamps. After all, such a combination is the old standard that has been used for years. The lamps are F96T12/CW (1500 milliampere) fluorescents and 100W, 130V incandescents in a 3:1 ratio by installed watts. The irradiance is fairly uniform (350 $\mu E\ m^{-2}\ s^{-1}$, ±5%) in the plant growing area as measured with a PPFD meter. Should the investigator be concerned about whether there is perfect spectral blending of the components of the fluorescent and incandescent irradiation in the plant growing area? To ensure that certain plants are not receiving a larger dose of incandescent irradiation than others, measurements can be made for adjustment of lamp and irradiance balance. The irradiance uniformity from either the incandescent or fluorescent component can be determined by measuring with a radiometer (300-2800 nm) in various locations in the growing area with only the fluorescent component operating, and then with only the incandescent component operating. Confirming that there is a uniform spectral blending (300-800 nm) from the lamp components requires measurements with a spectroradiometer.

The blending of the radiant output of lamps probably is becoming increasingly complicated with the combinations currently used. It is hoped that suitable optical systems are being used or designed to adequately blend the irradiation from lamps so that there is good uniformity of spectral radiance in the plant growing area.

Combinations of high intensity discharge (HID) lamps, or HID and incandescent lamps are common today. Some combinations include metal halide and tungsten halogen (incandescent), and metal halide and high pressure sodium lamps. The only way to be certain that spectral irradiance is uniform throughout the plant growing area is to measure it and make whatever adjustments are necessary to correct the optical blending of radiation.

In addition to the difficulty of optical blending with HID lamps there is the problem of manufacturers' variations in both lamps and ballasts. These variations are evident when examination of metal halide lamp systems shows difference in lamp color. Because variations in current supplied to lamps by the

ballasts can result in different arc temperatures in metal
halide lamps, there may be differences in spectral irradiance
from one source to another. This problem may be further
complicated by variations in lamp manufacturing. With burning
time it is also possible that certain metallic additives in the
arc of metal halide lamps may selectively condense out, thus
changing the spectral exitance. In high pressure sodium lamps,
variations in current supplied by the ballast also affect the
arc temperature, and thus the sodium vapor pressure in the arc.
The higher the current, the higher is the sodium vapor pressure,
resulting in emission of a greater proportion of spectral energy
in the longer wavelengths (600-700 nm). Lowering of the current
lowers the sodium vapor pressure and shifts the predominant
radiation to yellow-orange (560-630 nm). Manufacturers of lamps
and ballasts have established tolerance standards to minimize
these effects, but at times the extremes of ballasts and lamps
are combined in a system and make us visibly aware of these
variations.

In addition, the HID lamp system generally uses fewer lamps
per unit area than the fluorescent lamp system so that the HID
radiation exitance variations do not average out as well as in
the fluorescent system. The advantages of the HID lamp system
are energy saving and the higher irradiance levels than can be
achieved with the fluorescent-incandescent system.

Is one measurement at the beginning of an experiment enough?
It would be infinitely better to confirm that the irradiation,
spectral irradiance or photoperiod do not change during an
experiment. Even relatively short experiments may require daily
physical or visual monitoring. Timers do not always reliably
indicate whether lamps are turned on or off, or whether lamps go
on when turned on, and even whether plants are growing with the
appropriate photoperiod. A simple photocell monitor-record
would help here.

Framingham State College
Framingham, Massachusetts

The measurement guidelines for radiation proposed by the
American Society for Horticultural Science and the American
Society of Agricultural Engineers are to be commended. While
these measurements may make life more difficult and more costly
for some investigators, they will ultimately increase the
accuracy of research.

We have spent most of our effort discussing general measurements in general types of controlled environments. What about
spectral irradiance measurements in special or modified controlled environments? In such controlled environments there
may be special irradiation filtering systems; or single or
multiple monochromators that may be used to produce irradiance
in single or multiple narrow wavelength bands; or filters may
be used to subtract narrow wavelength bands from a broad
spectrum source of irradiation. For some investigations the
general types of controlled environments are being modified to
produce higher or lower levels of ultraviolet radiation: UV-A
(320-400 nm), UV-B (280-320 nm), UV-C (shorter than 290 nm),
red (600-700 nm), far-red (700-800 nm) or infra-red (above
800 nm) for stress or studies of plant morphogenesis. Such
measurements usually require the use of some kind of spectroradiometer or special radiometer.

Instrument calibration is very important for accurate
measurements. The frequency of necessary calibration will depend
upon the instrument. Calibration is perhaps best done by the
manufacturer, but there should be some way for the investigator
to check the need for recalibration. The manufacturer may be
helpful in supplying information or a means of using incandescent or special lamps to check the instruments and establish
that recalibration is needed.

Finally, it is well to reemphasize that measurement of
radiation is not simple or easy. It is an art and science in
itself and comes with the territory of radiation use.

REFERENCES

ASHS. Special Committee on Growth Chamber Environments. (1977).
    Revised guidelines for reporting studies in controlled
    environment chambers. *Hortscience 12*, 309-310.
Bickford, E. D., and Dunn, S. (1972). "Lighting for Plant
    Growth." Kent State University Press, Kent, Ohio.
Downs, R. J. (1975). "Controlled Environments for Plant Re-
    search". Columbia University Press, New York.
McCree, K. J. (1966). A solarimeter for measuring photosyn-
    thetically active radiation. *Agric. Meteorol. 3*,353-366.

RADIATION: GUIDELINES

J. Craig McFarlane

Monitoring and Support Laboratory
Environmental Protection Agency
Las Vegas, Nevada

When considering the radiation that stimulates photosynthesis
and other plant functions, the investigator typically thinks in
terms of light and subsequently of measuring light intensity and
quality. Strictly speaking, the use of the word "light" is
improper since light defines the electromagnetic radiation sensed
by human vision. Early measuring devices used filters to
simulate the same response as the human eye and were termed light
meters, with units of lux or foot-candles. Since most plant
responses are stimulated by the same wavelengths of radiation as
those we see, many light meter readings were taken and reported
in the early literature. In plants, several different light
absorbing pigments exist, each with its own absorption and action
spectrum. Therefore, various plant functions respond to different
spectral bands of light with different efficiencies. It is
obvious that plant pigments are not similar to the pigments
responsible for vision; therefore, measurement of light for
plant studies is a problem not solved by the concept of photometry as measured by foot-candles or lux. Abandoning photometry
in plant science has long been advocated by a host of scientists,
but journals still accept articles that report light meter

measurements. The committee recommendation is to cease the use
of photometry and consider it only when it is important to make
a historical comparison.

The concept of physiologically available irradiation (PAI)
was suggested in 1953 by the Committee on Plant Irradiation of
the Netherlands. A more recent proposal (Federer and Tanner,
1966) called for measurement of the number of incident photons
between 400 and 700 nanometers (nm). This measurement was termed
photosynthetically active radiation (PAR). The Crop Science
Society of America recommended (Shibles, 1976) that this measurement be designated officially as photosynthetic photon flux
density (PPFD). This is a descriptive title which accurately
indicates what is being measured. In the PAR and PPFD designa-
tions there is an inherent assumption that photosynthesis is
stimulated equally by all photons within the specified spectral
region (400 to 700 nm). This assumption was shown to be a good
approximation but does not account for differences in photosynthetic efficiencies at different wavelengths. Thus the novice
might incorrectly assume that this measurement is sufficient to
give exact correlation at all times with photosynthetic rates.
The units of PPFD have been typically expressed in terms of micro
Einsteins per square meter per second ($\mu E\ m^{-2}\ s^{-1}$). Incoll,
Long, and Ashmore, 1977, argued that the use of the Einstein
unit (E) has been incorrectly applied. They suggest that a more
correct and more usable unit for PPFD is moles of photons per
square meter per second ($mol\ m^{-2}\ s^{-1}$). This does not require
number changes from $\mu E\ m^{-2}\ s^{-1}$ but results in a more accurate
description of the units measured. The opposing argument states
that an Einstein is indeed defined as one mole of photons and
therefore there is no need to change. It seems that, since there
appears to be some confusion, the best solution is to simply go
to the most basic terminology. That will require that the definition as previously stated be used and what is actually measured

be reported: moles of photons per square meter per second. Since
the concept of measurement is PPFD, the appropriate units would
simply be mol m^{-2} s^{-1}. Despite the problem in expressing units
the concept of PPFD is of great value and is recommended as the
primary measurement of electromagnetic radiation for plants in
controlled environments.

Since instruments are not uniformly available for PPFD
measurements, the concept of photosynthetic irradiance (PI) has
received some attention. McCree discussed the concept of photon
absorption and compared photon flux measurements to irradiance
measurements. He stated that PPFD consistently gave a better
correlation to photosynthesis than PI. Therefore PI is recommended as an acceptable measurement when PPFD cannot be determined.

There is ample evidence that plants respond to wavelengths
beyond the region of 400 to 700 nm. Responses to ultraviolet
and infrared radiation have been identified and are currently
receiving the attention of many investigators. Therefore it is
suggested that total irradiance between 280 and 2800 nm be used
as the best second measurement to be taken. The preferred units
are watts per square meter (W m^{-2}).

None of the above measurements totally describes the electromagnetic radiant environment of the plants and all are therefore
inadequate for interpreting or understanding some critical studies.
Some investigations require specific details regarding narrow
spectral bands which stimulate morphogenic responses. Researchers
in these areas will, of course, measure the necessary wavelengths.
But instruments of known calibration and low price are not
available. We, therefore, encourage but do not recommend these
measurements. The very best measurement of the total radiation
environment is obviously the most complete one. Spectral
measurements are, however, difficult and costly to make. Those
conducting critical studies in which wavelength is an important
part of the controlled or uncontrolled environment will obviously
require spectral information. Spectral irradiance measurements

should cover between 280 and 2800 nm in 20 nm or smaller band
widths. These may be reported in a graph or table using the
dimensions of irradiance or photon flux density.

After the decision has been made on what to measure and in
which units to report, two critical decisions remain: where and
when to take the measurements. In controlled environments the
radiant environment of the plants continually changes. The cycles
of the refrigeration system cause the radiation levels to cycle
in concert with cooling and heating. Variations also occur which
are caused by changes in line voltage. These short time variations are not easily determined and many of them are uncontrolla-
ble. Thus, even though they are important in understanding many
studies, in general, their documentation was considered too
problematic to recommend as a guideline. Additional changes in
radiant intensity occur as lamps age. As plants mature in many
growth chambers, they alter their own radiant environment by
growing closer to the light source and creating shading for the
lower portions of the plants. In addition, light intensities
are not uniform across the top of the growing area. Differences
occur because of proximity to the walls of the chamber and
because of lamp positioning. It is realized that optimum information would be to have a measurement of the radiation to which
each leaf of each plant is exposed. This is clearly impossible
and some compromise must be used. To standardize procedures and
provide guidelines for the minimum acceptable data, it is
suggested that radiation measurements be taken at the top of the
plant canopy and that a mean plus a maximum and minimum value
be obtained over the plant-growing area. These measurements
should be taken at the start and finish of each study and
biweekly if the study extends beyond 14 days. Since spectral
irradiance is more complicated and difficult to obtain and interpret, it is suggested that this measurement be taken at the
center and top of the plant canopy at the beginning of each study.

If spectral distribution is important in a study, it should be remembered that each lamp type decays in its own peculiar way. Therefore, photon flux density or total irradiance measurement should be taken with each lamp type throughout the study period. This will allow the recreation of the spectral irradiance at any time, if that is needed.

Photoperiod is critical to most plant studies and obviously needs to be reported. Since some experiments use periods other than a normal 24-hour cycle, it is suggested that both the light and dark period be specified; for example, a statement such as: A 16/8 hour, light/dark cycle was used.

REFERENCES

Committee on Plant Irradiation of the Nederlandse Stichting voor Verlechtingskunde (1953). Specification of radiant flux density in irradiation of plants with artificial light. *J. Hort. Sci. 28*, 177-184.

Federer, C. A., and Tanner, C. B. (1966). Sensors for measuring light available for photosynthesis. *Ecology 47*, 654-657.

Incoll, L. D., Long, S. P., and Ashmore, M. R. (1977). SI Units in Plant Science. Commentaries in Plant Science *28*, 331-343.

Shibles, R. (1976). Committee Report. Terminology pertaining to photosynthesis. *Crop Sci. 16*, 437-439.

RADIATION: DISCUSSION

KLUETER: If we abandon photometry, how do we take advantage
of the voluminous data reported in foot candles or lux? Do we
disregard it or is there some way of making comparison?

McCREE: Obviously historical comparisons will require some
photometric measurements but my point was that this is a bad way
to proceed now. It is not generally possible to make accurate
comparisons or corrections between instrument types.

JAFFE: Photosynthetic and photomorphogenic light measurements
are important but there are other regions which are also important, i.e., blue, green, and infrared above 900 nm all stimulate
specific effects in some plants. Will a measure of the total irradiance give us the answer needed?

McCREE: When an investigator is doing photobiology he will
want to take spectral measurements. However, everyone should not
be required to do that.

KRIZEK: I would like to urge investigators not to ignore the
short and long portions of the spectrum. I suspect that there
are many photomorphogenic responses which have been overlooked by
not measuring irradiation outside the 400-700 nm band. We have
noticed, for example, that cotton can develop a red pigment in
response to radiation stress imposed from normal fluorescent
lamps. We now know that there is a considerable amount of UV-B
radiation in these sources and in some cases even a small amount
of germicidal ultraviolet radiation. Also some responses cannot
be accounted for by expressing radiation as square band pulses in
10 nm band widths. In view of recent advances that have been
made in the development of improved spectroradiometers, I urge
that spectral data be obtained for each nanometer where possible
and that those measurements then be integrated over the desired
band width of particular interest.

Karl Norris and his staff at the USDA Instrumentation Research Laboratory at Beltsville have developed an automated spectroradiometer that enables one to obtain weighted and unweighted spectral data for each nanometer from 250 to 400 nm. They have developed a three-scan instrument with capability of obtaining spectral data in the visible and near-infrared portion of the spectrum as well as in the UV region. Three manufacturers are now marketing instruments similar to these: EG&G (Salem, Massachusetts), Gamma Scientific (San Diego, California), and Optronic Laboratories Inc. (Silver Spring, Md.).

Consideration should be given to expressing spectral data on a logarithmic basis. Linear plots do not elucidate some spectral bands especially in the short wavelengths. Some portions of the spectrum may be very important biologically but may be overlooked if spectral data are presented on a linear basis.

KOSTKOWSKI: To make acurate spectral radiometric measurements the cost is probably 10 times as much as making broad (spectral) band measurements.

TISCHNER: In American literature PAR is used between 400-700 nm. In Europe 380-740 nm is used. In electrical engineering the best wave band to use is 380-740 nm because it is the visible range. I do not ask for a solution but I do think these differences should be recognized.

SKOOG: How much of the absorption of phytochrome is in regions other than the 660-730 wavelengths?

DOWNS: The main thing controlling some plant morphogenic responses is the amount of the far-red absorbing form of phytochrome present. This is controlled mainly by the ratio of red to far red in the light source. Blue light is not nearly as important since the blue ratios tend to nullify each other. Therefore, if one knows the ratio of red to far red, conversions are available to give the phytochrome ratio. Knowing that, plant responses can be predicted within reasonable limits.

WENT: So far all the discussion has been about short wave radiation. Long wavelengths are also very important, i.e., 10,000 nm. This radiation will probably have a big effect on morphogenetic and other processes. They are not considered at all and I recommend that measurements of long wavelength radiation should be included to adequately describe the conditions of plant growth.

McCREE: The long wave radiation primarily affects the plant's temperature. This could best be handled by measuring leaf temperatures.

WENT: I believe there are effects other than those of temperature from the very long wave lengths. We are currently investigating this by the use of an artificial sky radiator which reproduces the normal radiation sink effective in nature. Much more work needs to be done.

DOWNS: We recently had problems reminiscent of those to which Went refers. The symptoms were as follows. Geraniums grown under very high radiant flux densities died. Temperature measurements of leaf, substrate, etc. showed a high of 32°C in the substrate. When grown at lower radiant flux with substrate at 32°C, there was no problem. They did not grow well but they did not die. When a 10 cm water bath was placed between the lamps, the plants grew vigorously. This suggests some type of long wave energy effect that we are not measuring as heat.

KOLLER: In our phytotron in Israel we installed in the dark rooms a heating element producing the same wattage and in the same position as the lamps in the light rooms. This produces in the dark a comparable radiant load (in the long wavelengths) as do the lamps in the light.

CURRY: Question to McCree. Why did you select the wave band 300-3000?

McCREE: The wave band I reported was a rounded approximation. It was not intended to represent precise limits.

CURRY: The North Central Regional Committee which sponsored this meeting is preparing an instrument package which includes a radiation sensor. This will be passed around to cooperating laboratories within the region to verify the calibration of our own instruments. This is not intended to serve as an absolute standard but as a comparison to signal when meters need calibration. Calibration of the instruments in this package will be checked before and after each trip to a cooperating laboratory. We think this common circulating set of sensors may provide some coordination of data within the North Central region. If this works out perhaps others will be interested.

TIBBITTS: Physicists define quantum as the amount of energy carried by a photon. Thus if photobiologists utilize a quantum meter for measuring photosynthetically active radiation are they monitoring the energy or the number of photons of photosynthetically active radiation?

McCREE: Although physicists consider that a quantum is the amount of energy carried by a photon, photobiologists accept quanta to be interchangeable with photons and use quanta to refer to the number of particles or photons of radiation. Thus quantum yield is in moles per mole of photons or moles per Einstein and true energy of a mole of photons is proportional to the frequency or wavelength.

TANNER: I agree that we should continue the use of the term Einstein to describe 1 mole of photons. There is historical precedent and there appears to be no valid reason to abandon it.

KOLLER: Photobiologists now use the terms fluence and the fluence rates. Could McCree comment regarding this terminology?

McCREE: Fluence is an integrated value equivalent to energy. Perhaps the photobiologist may need these expressions but I believe that we are best served by continuing the use of flux measurements as already adopted.

McFARLANE: There seems to be a problem based on two alternate definitions of an Einstein, one describing a mole of photons, the other the energy of a mole of photons. These are clearly different concepts which are not comparable unless you know the

spectral distribution of the measured radiation. It seems clear
that a simple way to avoid this ambiguity would be to go the basic
unit and report what is measured, i.e. a mole of photons rather
than depend on knowing a definition.

TANNER: I don't think there is sufficient ambiguity to cause
any problem. How many here came into this meeting thinking that
an Einstein was a measure of energy rather than the number of photons? I think most people thought it was the number of photons.

McFARLANE: I would argue that there are many who work in
plant biology and report in terms of Einsteins and don't know what
Einsteins are. They have depended on others to specify the correct units for measuring radiant levels and have never gone beyond
the concept that it was a "light" value. The more obstacles we
can remove which require memorized trivia the more lucid will be
our results. Calling a mole of photons a mole of photons will allow that increased clarity we are all seeking. This argument is
reminiscent of those that allowed the use of diffusion pressure
deficit terminology. Such use confused every student who ever
tried to study plant water relations. The concept of water potential was a simplification easily understood and, when applied,
speeded the advancement of knowledge of plant water relations. I
argue for simplification in this case. Avoid confusion in spite
of historical precedence.

KOSTKOWSKI: The unit for photon flux in reciprocal seconds is
confusing. Hence use of mole reciprocal seconds is confusing.

In our self study manual we use photon reciprocal seconds to
avoid confusion even though it is contrary to normal SI usage.
Thus mole photon seconds^{-1} might be used for photon flux if the
mole rather than the Einstein is recommended.

McCREE: I was very pleasantly surprised with the recommendations for radiation guidelines, and I agree with almost everything
that was recommended. As I discussed earlier, however, I think
we should retain the Einstein. The confusion is with the photo-
chemists and not with photobiologists who have consistently used

the definition of 1 Einstein = 1 mole of photons. Although everyone doesn't know what an Einstein is, I think using the mole terminology would be confused with a mole of molecules and would add
confusion rather than clarity.

We should avoid using the term intensity which refers to a
source not a receiver. A non-specific term which I use for light
received at a surface is radiation level or light level.

McFARLANE: There is a real problem with abandoning the term
light. Strictly speaking, we really need to talk and write about
electromagnetic radiation, but practically this will be difficult
since we see light and we think in terms of light.

KRIZEK: In view of the fact that changes are being suggested,
now is the time to stop using the term *light*. I suggest adopting
the term visible radiation instead. Light implies eye sensitivity
responses. If we are going to divide the electromagnetic spectrum
into appropriate portions, I think it is time to handle them in
terms of radiometric concepts.

KOSTKOWSKI: Several people have talked to me recently about
measuring radiation in three bands, i.e., 400-500, 500-600, and
600-700 nm. If this could be done at a reasonable price would
this be desirable to plant scientists?

McFARLANE: I would say no. Measurements in three spectral
bands represent more information than in only one band but the interpretation is not clear. Some specific studies require detailed
measurements, but for a general guideline I think we must stick
with something immediately meaningful.

McCREE: These suggestions for three bands were generally
based on the Hoover photosynthetic action spectrum which had just
been released. It had peaks in the blue, red and a dip in the
green. We now recognize that the Hoover curve was not correct.
This is an additional reason for not wanting these bands.

WENT: We don't know enough about the effects of different
spectral regions. Therefore the more information we get, specifically in the green where I think there are inhibitory effects of

high intensities, the better we will know these effects.   Plants
do more than photosynthesize.   Light also stimulates morphogenic
effects.   We should therefore keep the measured values separate
so that when we learn what the effects of green light are we will
be able to go back through the literature and evaluate those responses.

TIBBITTS: Isn't there real difficulty in obtaining sensors
with uniform and sharp cut off filters at 400, 500, 600, and 700
nm?   Would there not be significant inaccuracies in band measurements under lamps with line spectra?   It seems that the lack of
accuracy would make the measurements meaningless.

WENT: I don't think so.   I think instruments could be made
very easily.

FORRESTER: Other industries keep standard filters which they
check with a spectral radiometer.   If that were done, inexpensive
instruments could be made using a combination of some inexpensive
plastic filters.

TANNER: A 3-division sensor will obviously not be needed if
spectral irradiance is available.   But if spectral irradiance is
not available a multi-band sensor (3 band) might serve as a diag-
nostic tool.

BERRY: Recommendations should reflect what should be measured in all experiments.   Special measurements should be left to
special needs.

McFARLANE: Let me please reiterate the North Central Regional
Committee position.   Those researchers specifically needing spectral information will make those measurements because that will
be part of their study.   The recommended standard is for some
measurement that will be generally taken and will allow comparison of procedures.   The committee recommendation was for general
use and not to elucidate specific details.

BATES: I would like to address the problem of where to measure.   If plants grow more than a few cm tall, measured radiation
levels at the top of the canopy will increase because of proximity

to the lamps. It is therefore important to decide what we are
trying to measure. Is it what plants are exposed to or is it
whether the lamps are changing over a period of time?

McFARLANE: Before trying to respond to the question of where
to measure, let me try to clarify a point about the suggested measurements. The guidelines you have are listed in the order considered most significant. Measurements at the top of the section
being the most important and those on the bottom the least important. The problem of where to measure is a difficult one. Our
attempt was to make the best suggestion for meaningful plant experiments. The top of the canopy represents a portion of the
plant. It is intended to measure what the plant is exposed to
rather than what the lamps are producing.

BICKFORD: Regardless of what measurement units are being reported, full description of the lamp source should be required.
This should include the proportions of each type lamp, allowing
for some understanding of the spectral distribution being reported. This will also allow another investigator to recreate similar conditions for study.

TIBBITTS: Is there a need to specify the lamp manufacturer?

BICKFORD: In some cases that may be necessary but generally
it is more important to specify the lamp combination.

KOSTKOWSKI: From the viewpoint of a radiometrist, I agree
completely with that idea. Also, whenever possible separate measurements should be made of each different lamp type in PPFD or
PI.

SALISBURY: Regarding the question of what we do with all the
data, I think the obvious answer is that we should keep data for
the future so that when anomolous results are recognized in future research we will have a background of information to rely on.

SEARLS: I suggest that it is important to include information
about the presence of and material of a transparent barrier between the lamp and the plants. This information, along with lamp

type will help define what UV radiation may be present in a growth
chamber.

KAUFMANN: I support the concept of keeping the suggested
guidelines as simple as possible while providing adequate information. Otherwise manuscripts could become mostly methods and be
too cumbersome for acceptance.

TIBBITTS: In his discussion McCree suggested that a ratio of
red to far red radiation may be a valuable measurement and that an
instrument could be developed to accomplish this. The fact that
some lamps have very narrow bands and sharp cutoffs causes concern about the usefulness of sharp cutoff filters in measuring
this red-far red radiation. What spectral band width would be adequate to monitor this ratio so that radiation peaks which may be
on the edge of either the red or far red response curves will not
be missed?

McCREE: Before an adequate answer can be given, more research
is needed on plant morphogenic responses. Without that complete
information I would estimate that 20 nm would be an adequate band
width.

KOLLER: At the Photobiology symposium last summer in Aarhus,
Denmark, Harry Smith of the University of Nottingham demonstrated
an instrument he had developed for measurement in the field of the
ratio of red to far red (660nm/730nm). This would be equivalent
to what would be absorbed by phytochrome in the red and far-red
absorbing forms. That instrument should now be commercially a-
vailable. And, as I remember, the price was to be in the range of
$200. He showed that with the same PAR the ratio of 660-730 de-
termined almost linearly the length and leaf production of tree
seedlings. He showed that there was a correlation between this
ratio and abundance of the two forms of phytochrome in the plants.
This measurement is meaningful in the field where a continuous
spectrum is present. A potential problem exists under artificial
light with a different spectral energy distribution. The ratio
may become meaningless and the results difficult to interpret.

In other words, under such conditions the measured ratio at 630 nm /730 nm may have little relationship to the level of active phytochrome in the plant, and therefore to its physiological response.

KOSTKOWSKI: The questions concerning measurement of phytochrome might best be handled with spectral measurements. There are some very small, inexpensive ($200-$300) monochromators currently becoming available which have a very wide exit slit and a narrow entrance slit. With an appropriate grating and a detector I think it would be possible to make a device for obtaining the data for $500 to $600.

KLUETER: The guidelines as originally stated included 2 levels of priorities. One which was required and the second was to be optional. What has happened to that ordering?

McFARLANE: That was the intent of the guidelines as presented. PPFD was recommended as the primary data with other measurements to be included as needed.

KRIZEK: I would like to point out that some of the questions raised regarding lamp loading, barrier, etc., have been covered in the American Society for Horticultural Science (ASHS) guidelines published in Hortscience (1970, 1972, 1978), which served as a starting point for these recommendations. I want to comment again regarding the term "light". In the visible portion of the spectrum this is not a serious problem but in the UV or IR regions some people often speak of UV light or IR light. To do so is obviously incorrect. Therefore, although it may be more convenient to speak of light, to be accurate and consistent we should use the term radiation or radiant energy and speak of lamps as UV, visible or IR radiation sources. Similarily, the term radiation measurement should be used instead of light measurement.

PALLAS: I do not agree with the suggestion of taking measurements at the top of a canopy. Often we have dissimilar plants in a chamber and this would cause much confusion. We take our measurements without plants at the soil surface and report the plant

height. Measurement at the top of the canopy would cause confusion with which I do not know how to deal.

KRIZEK: I suggest that even in a mixed canopy we should still
measure radiant energy at the top. Most critical photomorphogenic
responses take place at the apical meristems. This seems to be
the most logical region to represent the radiant energy flux to
which a plant is exposed. Other measurements are encouraged but
measurement at the top of the plant canopy is suggested as the
minimum requirement.

ANDERSON: From an operational point of view, maintenance of
a light level requires the specification of a measuring point.
Since plants cause shading and affect the reflection from the side
walls, the top of a plant canopy seems a reasonable point to
specify.

If the response of a particular sensor is known, the quickest
and most reasonable way to obtain an accounting of the spectral
distribution in a chamber is to specify lamp types and the mix
used.

KOSTKOWSKI: I would also support the idea that lamp description is the best description of spectral distribution short of a
precise spectral radiometric analysis.

McFARLANE: The committee spent much time on where to measure
radiation levels. This question can be put into perspective by
asking: "Do you want to know what is happening to the lamps or
what is happening to the plants?" Engineers may need to know what
is going on with the lamps. Certainly to effect controls and monitor chamber performance some measurement of lamp performance will
be needed. On the other hand plant scientists need to know the
conditions that affect the plants they are studying. In a cooper-
ative study of the Growth Chamber Committee of ASHS we required a
certain radiation level at the start of the study and then let the
lamps decay as they would and measured the radiation as it changed
due to lamp decay and plant growth. I think our concern should
be for the plant environment and we should adopt the top of the

canopy as representing a reasonable point for collection of minimum data.

CURRY: I support the top of canopy concept. Most of the photosynthesis occurs there and most interest is in that value. I want to comment about reporting methods. If an investigator cannot find out what was done by reading the methods section, the rest of the paper is not worth considering. In looking back on data, sometimes the details of the methods turn out to be the most important aspects of the research.

MILLER: Using vertical mirrors on the walls alongside the plants, gives essentially the same result as having an infinitely large field of plants under a uniformly bright sky. Although it may not be practical to use mirrors, aluminum foil may be used on the walls with the shiny side out. I think this is a good way to approach the problem of variation in light level with plant growth.

I would also like to speak about the arguments that have been going on here about choices of units. Investigators seem always to be trying to find the ultimate in units and of course the best means of measurement in these units. This is unlikely to be achieved because development means relentless change. New experiments will require new methods of expression and of performing measurements. I feel that it is important for investigators to at least state in the papers what they do know about their system for radiant measurement, including lamp type, sensor type, etc.

HELLMERS: At Palmerston North, New Zealand they have done as Miller suggested. They used plate mirrors in their chambers. At the North Carolina State and Duke University phytotrons we have more than 100 chambers with mirrorized walls with solid light ceilings and they work very well.

PALLAS: Alzak gives a near mirror effect and many of us have gone to these reflective surfaces. However, this does not eliminate the problem of where to measure. Measurement at the top of the canopy in some cases means the top of an inflorescence.

Radiation at that point does not necessarily represent a level indicative of a plant response. A more integrated value appears to be warranted involving measurements at the soil surface and again at the topmost layer of leaves that account for about 80% of photosynthesis.

TIBBITTS: I want to respond to Pallas' suggestion about measuring somewhere down in the canopy. I see 2 philosophies for growing plants in growth chambers: (1) Try to get a solid canopy to duplicate somewhat the competition found in field conditions, and (2) Space plants out so there is no competition among them. How would you reconcile the measurements in those 2 situations?

PALLAS: If the investigator is doing something relevant to the field he would probably work with a canopy and could probably get by with one measurement. In a sparsely populated chamber it might be more important to make 2 measurements because reflective properties of the growth chamber may be more important.

BICKFORD: I would like to hear comments from growth chamber manufacturers regarding the radiation level specifications in their sales literature.

RULE: I would like to compliment the committee on doing an excellent job in making recommendations. Many of us still use foot candles in our literature and I think we should start using PAR. We will be happy to give our values that way.

SEARLS: We are in the middle of changing all our literature into metric units. For us, now is the time to make other changes. What you want is how we will reflect it.

TAYLOR: The user dictates what we will put in our literature. We are just now printing ours in µE.

CURRY: Could we have some more reactions regarding the units used for measurement and reporting?

NORTON: I suggest a vote be taken to show favor for different units, specifically the units Einsteins or moles of photons. (Editors Note: The vote indicated that the large majority preferred Einsteins).

TIBBITTS: We would appreciate reaction over the need for specifying the input wattage of fluorescent and incandescent lamps.

READ: If you are concerned with commercial application, input wattage allows for an economic evaluation. In our application that information is very important.

PALLAS: What relevance is there in wattage when manufacturers may claim improved efficiency such as with power-groove lamps compared to other types of fluorescent lamps.

BICKFORD: Reporting wattage would be important in describing radiation from a light bank of mixed lamp types because of the different contributions of various light sources. For each light source used the spectral radiation would generally be proportional to the lamp wattage. Thus it would be desirable to report the number and wattage of each of the lamp types used.

Knowing the electrical demand is an important consideration for power requirement, i.e., cost. Efficacy is generally expressed as lumens/watts. However some papers have reported PPFD/watts. Efficacy is rather standard between manufacturers for the same type of lamp, i.e., cool white 1500 milliamp fluorescent lamps are all similar.

KLUETER: New lamps are more uniform than those that have been in use for some time. The temperature at which a lamp runs also influences its efficacy. We should also consider when during the light period and refrigeration cycle to make measurements.

DOWNS: Historically the mixture of lamps started out with 10% of illuminance provided by incandescent lamps and the remaining from slim line type fluorescent lamps or from carbon arc lamps. If this ratio of illuminance is used with today's lamps the ratio will be different. According to Borthwick, 10% of incandescent illuminance was the only amount ever tried. This carryover of the traditional ratio of fluorescent to incandescent is therefore open to question. Some investigators suggest that there is not enough incandescent light to obtain normal plant growth and several researchers are currently trying to determine the ratio that might give the most representative plant growth.

TIBBITTS: I think Downs alluded to the important reason for specifying the imput wattage. With a mixed lamp source, specifying the wattage of each type gives some information about the balance of red to far red, as well as the level of infrared radiation being provided.

McFARLANE: The problem of light cycling with refrigeration cycling is one which most of us recognize but it is difficult to describe. The committee recommendation is to give the limits of that cycling but not necessarily to try to describe it. In the ASHS group we did evaluate the cycling during lamp warm up and again later in the light cycle. It turned out that we did not really know what to do with the data.

TEMPERATURE

Frank B. Salisbury

Plant Science Department
Utah State University
Logan, Utah

INTRODUCTION

Virtually everything in the universe is profoundly influenced by heat interactions. Growing plants are certainly no exception. We know that heat interactions depend on the relative kinetic activities of microparticles between two systems in thermal communication. Since energy is conserved as objects and their surroundings arrive at thermal equilibrium, the average translational kinetic energies of the different particles (atoms, molecules, ions, etc.) in the two communicating systems become the same. Thus, heavier molecules in a gas will have lower average velocities than lighter molecules, atoms, etc. Since the particles are continually colliding and exchanging kinetic energy (billions of times per second for each particle), few particles will actually have a velocity equal to the average. The velocities of similar particles in a gas will be distributed according to the Maxwell-Boltzmann equations (Fig. 1).

Temperature proves to be difficult to define in a rigorous and completely satisfying way, but for our purposes we may think of temperature as the condition of a body that determines the direction and rate of the heat interaction to or from other bodies. Because heat is energy transferred across a boundary in response

to temperature differences across that boundary, heat has the
same units as energy; namely, joules in the SI system, although
calories are still widely used. Since modern thermodynamics
defines heat only in terms of energy transfer, we should not
speak of heat content, only thermal energy content (see, for
example, Van Wylen and Sonntag, 1976, pp. 76-77).

The energy stored within a system depends upon the mass
or extent of the system; hence, thermal energy is an extensive
property of matter. Temperature is independent of the extent
of the substance; hence, temperature is an intensive property of
matter. In a rough sense, we can think of temperature as an
intensity or potential measurement of translational energy of
microparticles, analogous to concentration, pH, vapor pressure,

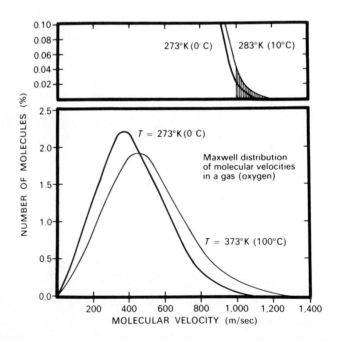

FIGURE 1.   The Maxwell-Boltzmann distribution of molecular
velocities in a gas at two temperatures 100°C apart.   The curves
at the top show the high-velocity portion for a gas at two
temperatures 10°C apart.   The area indicated by the vertical
lines can be seen to approximately double in going from the
lower to the higher temperature.   Temperature is a partial
expression of this translational energy.   (From Salisbury and
Ross, 1969.)

voltage, etc. The temperature difference between two bodies
suggests the potential for the heat interaction or what is
commonly called the heat transfer between them.

Temperature is detected by observing changes in the
properties of various materials as they reach thermal equilibrium
with their surroundings. The expansion of liquid in a mercury
or alcohol thermometer is calibrated to a graded scale with
certain reproducible points such as the equilibrium temperature
between melting ice and water. Expansion of metals, variations
in electrical resistance of various materials, the thermoelectric
production of a current, or emitted radiant energy are also used
to measure temperature.

PRINCIPLES OF HEAT TRANSFER

Consideration of temperature as it relates to growth of
plants in controlled environment facilities must be based upon
known principles of heat transfer between objects and their
environments. When the radiant energy load is relatively low
(say less than 150 W m^{-2} total, or 22 klx illuminance, which is
not meant to imply that these are equivalent), the relatively
small excess of radiant energy absorbed over that emitted by
the leaf causes only a small increase in leaf temperature, and
this is sufficient to dissipate the small positive net radiation
by conduction-convection between the leaf and the air moving
over it and by transpiration. As a result, leaf temperatures
and air temperatures remain in rather close agreement (e.g.,
within 1 or 2°C). Until relatively recently, this has been the
situation in most controlled environment facilities. It is
becoming economically feasible, however, to produce radiation at
solar levels in growth chambers. Under these conditions, leaf
temperatures may deviate by 10°C or more from air temperatures,
and the principles of heat transfer can no longer be ignored.

A brief review presenting the principal concepts and equations
is appropriate here (see reviews in Raschke, 1960; Gates, 1965;
Drake, Raschke, and Salisbury, 1970; Gates and Schmerl, 1975;
and Salisbury and Ross, 1969 and 1978).

Basic to the approach is the energy balance equation. It
states that energy flux by all the mechanisms of energy exchange
between an object and its environment must total zero (0). If a
term in the equation is positive, it means that the object is
absorbing a net balance of energy (represented by that term) from
its environment; if negative, it is losing a net balance of
energy to the environment. For a plant, the energy balance
equation may be written as:

$$0 = Q + H + V + B + M$$

where:

Q = net radiation. (Negative term = leaf is radiating more
    energy to its surroundings than it is absorbing from them;
    positive, leaf is absorbing more radiant energy.)

H = sensible heat transfer (includes conduction and
    convection).

V = latent heat of vaporization, the transpiration term.
    (Negative = water is vaporizing; positive = dew or frost
    is condensing.)

B = storage. (Positive = leaf temperature is increasing.)

M = metabolism and other factors (photosynthesis, respiration,
    water arriving from soil at some temperature other than
    that of the leaf, and other factors).

When leaf temperature is constant (steady-state conditions),
B is equal to zero. Factors constituting M are also usually
ignored because they are minor. Hence the equation may be
written:

$$0 = Q + H + V$$

Energies of all components are now best expressed in $W\ m^{-2}$, which are SI units, but in the older literature, including much work referred to in this paper, units are often $cal\ cm^{-2}\ min^{-1}$, which combination of units has been called the Langley. Gram calories are multiplied by 4.184 to give joules, and $W = J\ s^{-1}$. In general, several to-be-discarded-but-familiar units are used in this paper along with their SI equivalents.

The net radiation is equal to the radiant energy absorbed by the leaf minus the energy radiated from the leaf. In the light, there are two components to the absorbed energy: radiant energy emitted from the light source, and long-wave thermal radiation emitted by all objects in the plant's environment. In each case, the total energy falling on the surface of the leaf must be multiplied by an absorptivity coefficient that expresses the absorbed fraction of the energy striking the leaf. Absorptivity of a leaf depends on wavelength, but a single absorptivity coefficient can be determined for a given leaf irradiated by a given light source. Absorptivity coefficients (e') for different leaves irradiated by solar and artificial light sources vary widely, roughly from 0.44 to 0.88. The absorptivity of long-wavelength thermal radiation coming from objects in the plant's environment is much higher ($e'' \simeq 0.95$).

From a theoretical standpoint, it might be desirable to separate absorbed radiation into photosynthetically active radiation (PAR) plus all other wavelengths. Leaves have high absorptivities in the PAR region and low in the near-infrared wavelengths produced by most light sources, so absorptivities for PAR would be higher than those suggested above for source radiation. In practice, little would be achieved by artificially separating PAR from other, absorbed radiation in an energy transfer study, since only the total is important, and it is usually simple to separate visible sources from thermal radiation by turning off or shading the visible source.

Energy absorbed (Q_{abs}) may be expressed as follows:

$$Q_{abs} = e'Q_v + e''Q_{th}$$

where:

Q_v = total radiant energy from the sources that emit photosynthetically active radiation.

Q_{th} = total thermal radiant energy from all other sources.

Incidentally, it is not unusual even under conditions of bright
sunlight for the thermal radiation to equal or be greater than the
energy from sources that emit PAR or visible light.

The energy emitted by the leaf is a function of the fourth
power of the absolute temperature (the Stefan-Boltzmann law).
This emitted energy is subtracted from the absorbed energy to give
the net radiation:

$$Q = Q_{abs} - e''\sigma T^4$$

where:

e'' = emissivity or absorptivity of long wave radiation (about
0.95 for leaves at normal temperatures).

σ = the Stefan-Boltzmann constant (8.135 x 10^{-11} cal cm^{-2}
min^{-1} K^{-4}; 3.404 x 10^{-10} joules cm^{-2} min^{-1} K^{-4}; or
5.670 x 10^{-8} W m^{-2} K^{-4}).

T = absolute temperature (K).

Sensible energy transfer by convection (H) is proportional to
the temperature gradient (the difference in temperature between
the air and the leaf) and inversely proportional to the resistance
of the boundary layer:

$$H = \frac{(T_a - T_1)c_p \rho}{r_a} = \frac{\Delta T c_p \rho}{r_a}$$

where:

T_a = air temperature

T_1 = leaf temperature

r_a = boundary layer resistance. (Traditional units are s or min cm^{-1}; SI units are s m^{-1}.)

c_p = specific heat of dry air (0.239 cal g^{-1} $°C^{-1}$; 1.000 joules g^{-1} $°C^{-1}$).

ρ = density of air (g cm^{-3}; kg m^{-3}).

The values for c_p and ρ describe the ability of air to absorb thermal energy.

The reciprocal of the boundary layer resistance, including the specific heat and density terms for air, is called the convective transfer coefficient or the heat transfer coefficient (h_c):

$$h_c = \frac{c_p \rho}{r_a} \qquad \text{(cal } °C^{-1} cm^{-2} min^{-1}; \text{ J } °C^{-1} cm^{-2} min^{-1}; \text{ W } m^{-1} °C^{-1})$$

$$H = \frac{\Delta T c_p \rho}{r_a}$$

$$H = h_c \Delta T$$

The heat transfer coefficient is analogous to permeability and is a measure of thermal energy conductivity through the boundary layer. The higher the coefficient, the more rapid is the transfer of energy by convection. Thus a thin boundary layer is associated with a large heat transfer coefficient or a small boundary layer resistance. The heat transfer coefficient is often used by heating engineers instead of boundary layer resistance because it exhibits an almost linear relationship when it is plotted as a function of the square root of wind velocity (Fig. 2). In plant physiology, resistance is often used so that it can be summed with other diffusion resistance factors in the plant. The boundary layer thickness is a function of leaf size and shape and the type (laminar or turbulent) and velocity of air flow. Higher wind velocities and smaller leaves lead to thinner boundary layers, lowered resistance, and thus more rapid convec-

tive heat transfer. Figure 3 illustrates basic concepts of the
boundary layer.

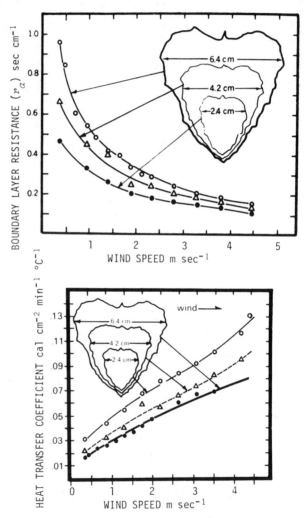

FIGURE 2. *Boundary layer resistance and heat transfer coeffi-*
cient, both shown as a function of wind speed over copper plated
leaves of Xanthium strumarium. Boundary layer resistance
decreases with increasing wind and with decreasing leaf size,
while the heat transfer coefficient, its reciprocal, increases
with increasing wind and smaller leaf size. The heat transfer
coefficient more closely approaches a linear function of wind
speed. (From Drake, Raschke, and Salisbury, 1970.)

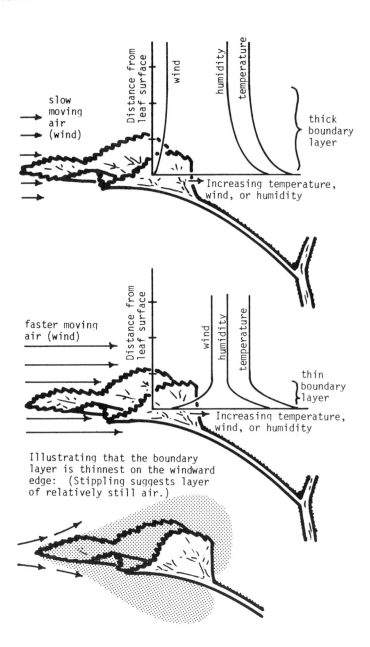

FIGURE 3. Some principles of the boundary layer and heat transfer by convection. The two upper leaves are assumed to be at the same temperature and in air at the same temperature; only wind speed is different. The thickness of the boundary layer decreases with increasing wind speed.

The equation for the latent heat of vaporization (the transpiration term) is closely analogous to the convection equation: Energy transferred is proportional to the driving force and inversely proportional to the resistance. In this case, the driving force is the difference between the vapor pressure in the leaf (p_l) and the vapor pressure in the atmosphere beyond the boundary layer (p_a). The resistance has two components: the boundary layer resistance (r_a) and the leaf resistance (r_l), the latter consisting primarily of the resistance to diffusion through the stomata. Thus the equation for the transpiration term (V) can be written as follows:

$$V = \frac{(p_a - p_l)c_p \rho}{\gamma(r_l + r_a)} = \frac{\Delta p \; c_p \rho}{\gamma(r_a + r_l)}$$

The equation also contains the specific heat of air (c_p) and the density of air (ρ), as in the equation for convection. In addition, it includes the psychrometric constant $(\gamma = 0.5$ mm Hg $°C^{-1}$; 0.0666 kilopascals $°C^{-1})$.

APPLYING THE PRINCIPLES OF HEAT TRANSFER

Several factors in a plant's environment influence its energy exchange process. These factors will be discussed in relation to how they affect leaf temperature, which is certainly of high metabolic significance to a photosynthesizing plant. Most of these factors can be modified by the way that growth chambers are designed and operated.

Species: Leaf Thickness and Transpiration Properties

The temperature response of a given plant to its environment will depend first of all on the plant itself (Fig. 4). Thicker leaves have a higher heat capacity on a unit area basis and thus

heat more slowly but reach higher temperatures above air
temperature under heavy radiation loads (Mellor, Salisbury,
and Raschke, 1964). There are also important physiological
differences, few of which have ever been carefully studied. A
cocklebur (*Xanthium strumarium*) leaf is almost identical in
appearance to a sunflower (*Helianthus annuus*) leaf. The cocklebur plant can only grow where it is supplied with ample amounts
of water (e.g., near ditch banks or in moist climates). At high
air temperatures, its leaves are cooled below the air temperature
by transpiration. On the other hand, sunflower plants grow on
dry, semi-desert hillsides in the arid west, where their leaves
reach temperatures considerably above air temperature.

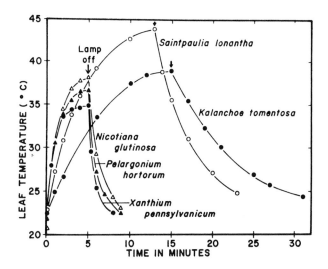

FIGURE 4.   Heating and cooling rates for various species.
Radiant energy = 1.34 cal cm^{-2} min^{-1}; air temperature = 22.3 to
25.5° C; relative humidity = 48%. (From Mellor, Salisbury, and
Raschke, 1964.)

Radiation Load

Of course, the difference between leaf and air temperature
will increase with increasing radiation absorbed by the leaf
(Fig. 5). This was illustrated in an esthetically pleasing way
one summer solstice at 14,100 ft (4300 m) on Mt. Evans in
Colorado (Salisbury and Spomer, 1964). The sky was perfectly
clear, and the sun was as high as it gets during the entire
year. Thermocouples were attached to the leaves of several
plants and their temperatures recorded at intervals throughout
the day (Fig. 6). The leaf of one rather succulent plant
(*Bessya*) reached a temperature about 22°C above that of the air.
The terrain was rocky, and when the shadow of a rock passed
over a leaf, sharply reducing the radiation load, leaf temperature
dropped to and below air temperature.

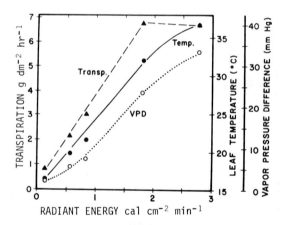

    *FIGURE 5. Transpiration and leaf temperature at various
radiant energy levels using Xanthium leaves in a growth chamber.
Note the close relationship to vapor pressure difference. Leaves
were damaged at the highest radiation levels. Air temperature =
15 ±0.5°C; relative humidity = 94 to 98%. (From Mellor, Salisbury,
and Raschke, 1964.)*

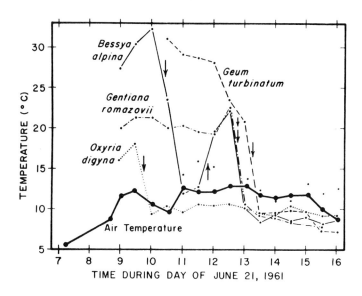

FIGURE 6. Some leaf temperatures of alpine plants in the
field. Thermocouples were inserted into leaves (or contacted
leaf surfaces) on Mount Evans, Colorado, June 21, 1961; skies
were clear. Arrows indicate when the shadow of a rock passed
over a given plant; in the case of Bessya alpina, the leaf was
illuminated for the second time just after noon. (From
Salisbury and Spomer, 1964.)

Air Temperature

Table 1 shows a number of measurements and calculations for
an experiment using single attached cocklebur leaves in a wind
tunnel (Drake, Raschke, and Salisbury, 1970). At the low air
temperature, leaf temperature was well above air temperature;
this situation was exactly reversed at the high air temperature
(and low humidity), where the leaf was being cooled by rapid
transpiration. At the high air temperature, the leaf was gaining
energy, not only from the incoming radiation, but also by convection from the surrounding warm air. All of this extremely high
heat load was being dissipated by transpiration.

TABLE 1. *Energy Exchanges of a Xanthium strumarium (Cocklebur) Leaf under Two Different Temperature Environments*

Experimental Conditions:

Air temperature (T_a)	5°C	40°C
Leaf temperature (T_l)	15.3°C	36°C
Approximate relative humidity	100%	48%
Vapor pressure difference (leaf - air)	6.49 mm Hg	18.0 mm Hg
Wind speed	90 cm s^{-1}	90 cm s^{-1}

Radiation Exchanges (cal cm^{-2} min^{-1}):

Xenon lamp radiation (filtered)	0.486	0.487
Xenon radiation absorbed by leaf (Q_i)	0.340	0.341
Radiation absorbed from surroundings (Q_s)	0.710	0.838
Radiation emitted by leaf (Q_l)	-0.545	-0.719
Net radiation $(Q = Q_i + Q_s + Q_l)$	0.505	0.460

All Energy Exchanges (expressed as cal cm^{-2} min^{-1}):

	5°C Air temp.		40°C Air temp.	
Energy lost from the leaf:	A[1]	B[2]	A[1]	B[2]
Radiation (Q_l)	-0.545	Ignore	-0.719	Ignore
Convection (H)	-0.310	-0.310	+0.120	+0.120
Transpiration (V)	-0.193	-0.193	-0.590	-0.590
Total lost	-1.048	-0.503	-1.189	-0.470
Energy gained by the leaf:				
From the xenon lamp (Q_v)	0.340		0.340	
From the surroundings (Q_s)	0.710		0.838	
Total absorbed	1.050		1.179	
Net radiation (Q)		0.505		0.460
Error (difference between total lost and total absorbed or Q)	0.002	0.002	-0.010	-0.010
Energy exchanges expressed as percentages:				
Radiation	-52.0%	----	-60.5%	----
Convection	-29.6%	-61.6%	+10.1%	+25.5%
Transpiration	-18.4%	-38.4%	-49.6%	-125.5%

[1]Calculations utilizing total radiant energy absorbed and emitted.
[2]Calculations utilizing net radiant energy exchange.

Wind Speed

Figure 7 is an extension of the data just discussed (Drake, Raschke, and Salisbury, 1970). It includes the results of many experiments utilizing different air temperatures and different wind velocities over leaves in the wind tunnel (fairly constant humidities). Because of the effect on the boundary layer, increasing wind decreases the difference between leaf and air temperature. For air temperatures below the cross-over point where leaf and air temperature are the same, increasing wind cools the leaf; for air temperatures above this cross-over point, increasing wind warms the leaf. Figure 8 shows that increasing wind may sharply reduce transpiration as it reduces leaf temperature (Mellor, Salisbury, and Raschke, 1964), a phenomenon that occurs when leaf resistance is high. All of this is described nicely by the equations already discussed here. The driving force for transpiration is the difference in vapor pressure inside the leaf and beyond the boundary layer. As leaf temperature decreases, internal vapor pressure drops sharply.

Does this imply that growth chambers should utilize high wind velocities? To an extent, yes, but it is important to realize that plants are highly sensitive to mechanical stresses. Shaking a tomato plant 10 seconds each day will reduce its growth to only 60% that of controls (Jaffe, 1973; Wheeler and Salisbury, 1979. The mechanical stresses can be applied by manual shaking, rubbing, spraying with water, applying wind, or in other ways (Neel and Harris, 1971; Victor and Vanderhoef, 1975; Jaffe, 1976; Mitchell, 1977; and Wheeler, 1978). To an extent, growth reduction by mechanical stress is probably desirable in growth chamber work, because greenhouse plants, protected to some extent from mechanical stresses, are elongated and leggy; they are certainly not typical of plants in most natural or agricultural situations. But the response to mechanical stresses can become extreme. Handling a cocklebur leaf each day, for example, not only

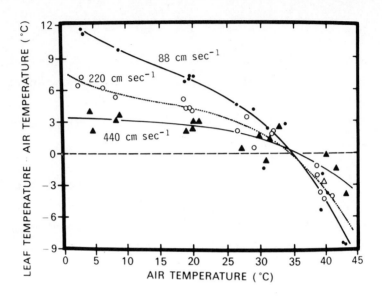

FIGURE 7. The difference between leaf temperatures of a Xanthium leaf and air temperatures for 3 wind velocities and as a function of air temperature. Curves are third order polynomials drawn by computer to match the data. At air temperatures below about 34°C, leaf temperatures are above air temperatures, but less so with increasing wind speeds; above about 34°C, leaf temperatures are below air temperatures, this being accounted for by extremely rapid transpiration rates (not shown). (From Drake, Raschke, and Salisbury, 1970).

reduced its growth to about 70% of controls but caused it to turn yellow and die (Salisbury, 1968). It is important, then, to arrive at some compromise for growth chamber wind velocity, realizing that wind is both an effective cooling agent and a mechanical stimulus.

Light Quality

The extent to which incoming radiation will influence leaf temperature depends strongly upon its quality or wavelength balance. This was demonstrated several years ago in an interesting experiment in which visible light with virtually all infrared radiation filtered out was used to illuminate plants

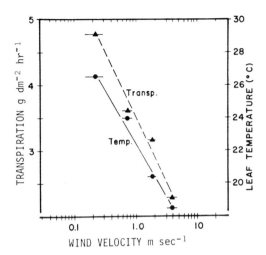

FIGURE 8. A striking example of decreasing transpiration with increasing wind speeds. The apparent cause is the equally sharp drop in leaf temperatures and hence the greatly lowered vapor pressure difference between the leaf's internal spaces and the atmosphere. Radiant energy = 1.0 cal cm^{-2} min^{-1}; air temperature = 15 ±0.5°C; relative humidity = 94 to 96%. (From Mellor, Salisbury, and Raschke, 1964.)

in one set of experiments, while visible light plus near-infrared radiation was used to illuminate plants in another set of experiments (Mellor, Salisbury, and Raschke, 1964). The results, shown in Figure 9, illustrate that visible light was much more effective in increasing leaf temperature than was visible plus near-infrared. We usually think of infrared radiation as a heat source, but *near*-infrared (about 700 to 1200 nm) is almost completely transmitted or reflected by thin leaves (90% or more). Removing near-infrared from growth chamber illumination sources might reduce the heat load in the growth chamber without greatly influencing plant growth, although this certainly needs experimentation. One advantage of the use of fluorescent lamps might be the virtual absence of near-infrared radiation from these lamps.

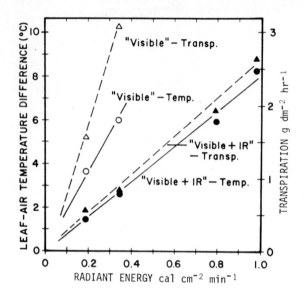

F.IGURE 9. Differences between leaf and air temperatures and
transpiration rates as a function of radiant energy load for
leaves illuminated by "visible" light only and for leaves
illuminated by "visible plus near-infrared radiation." "Visible"
light was obtained by filtering xenon radiation through copper
sulfate solutions. Leaf temperatures and transpiration rates
are lower when near-infrared radiation is part of the total
radiation load because leaves absorb very little of the near-
infrared part of the spectrum; leaves absorb much more of the
visible part of the spectrum, so that portion is more effective
at heating leaves. At 0.34 cal cm^{-2} min^{-1}, "visible" gave
readings of 7950 ft-c and "visible + IR" gave readings of only
2150 ft-c. Air temperature = 15 ±0.5°C; relative humidity =
86 to 90%. (From Mellor, Salisbury, and Raschke, 1964.)

Atmospheric Moisture

    Decreasing relative humidity (increasing vapor pressure
deficit between a leaf and the atmosphere) should increase
transpiration and thus reduce leaf temperature, and this has
been observed (Matsui and Eguchi, 1972 and 1973). Yet Lange,
et al. (1971) observed that stomata of *Polypodium vulgare* and
Valerianella locusta closed in response to low relative humidity

and opened in response to high humidity, thereby adjusting transpiration to be more-or-less constant at various humidities.

We can gain some perspective on how humidity might influence transpiration and thus leaf temperature by comparing its influence with that of leaf temperature itself. A leaf in air at 20°C and 90% relative humidity and at the same temperature as the air has a gradient (called the saturation deficit) of about 0.234 kPa from its internal spaces to the atmosphere beyond the boundary layer. Increasing the leaf temperature to 30°C without changing air temperature or humidity establishes a saturation deficit of about 2.14 kPa. To achieve about the same effect by lowering atmospheric moisture without changing leaf temperature, relative humidity must be reduced to about 8.7%. In spite of such theoretical considerations, effects of humidity on transpiration and leaf temperatures need much more study with many species and under a variety of conditions.

Other Complications

Several other factors might be expected to influence leaf temperature, including some that are controlled by the plant's physiology. Drake and Salisbury (1972) found that a 72-hour pretreatment at high (40°C day, 36°C night) or low (10°C day, 5°C night) temperature strongly modified subsequent leaf temperature and transpiration of cocklebur plants. Transpiration was considerably reduced and leaf temperatures elevated for plants that had received the low-temperature pretreatment compared to the high-temperature pretreated plants. This was true at air temperatures from 5 to 45°C.

Stomatal aperture has also been shown to exhibit an endogenous rhythm under constant conditions (Meidner and Mansfield, 1968). This controls transpiration rates and thus leaf temperatures.

MEASURING TEMPERATURE

Making good temperature measurements depends on suitable
application of heat transfer principles. When we speak of
temperature in a growth chamber, we usually think of air
temperature. To obtain a good measurement of air temperature, we
need to maximize convective heat transfer (in the absence of
evaporation, of course). Thus, our sensor must be shielded from
the radiation environment, preferably by some insulating
material that will not itself heat up in response to the
radiation. Perhaps an infinite variety of shields is possible,
but one that we devised many years ago from two plastic cups
(Fig. 10) seems to meet the above theoretical requirements
adequately. Sensors in some commercial chambers are placed in
aspirated tubes to speed up the equilibrium between sensor and
air temperature.

To give an accurate reflection of the rapid changes in air
temperature, the sensor should have a low heat capacity, although
it might sometimes be desirable to average out these fluctuations
somewhat - as the plant probably does - in which case a sensor
with a somewhat higher heat capacity may be desirable. Since a
thermocouple can be made with extremely fine wire (low heat
capacity), it may be the ideal temperature sensor. It is
especially well suited for multichannel recorders, but it may be
less suited as a source of input information to a control
system, since the signal is weak and must be electronically
amplified to operate control systems. Thermistors (in which
electrical resistance of a semiconductor changes with temperature)
may be well suited for certain relatively narrow temperature
ranges, but their nonlinearity of response makes them rather
difficult to work with. (These problems can be overcome with
suitable circuits.) Bimetallic strips and bulbs with liquid,
although both have a considerable heat capacity and thus lag in
response, are often used. For many applications they are quite

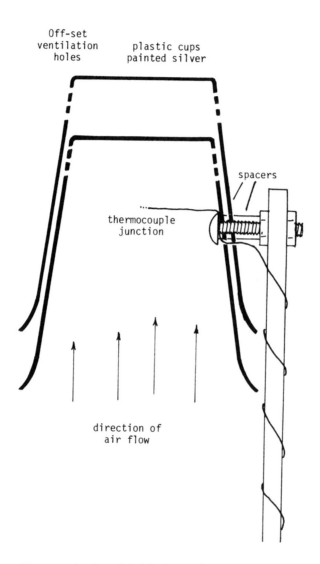

Off-set
ventilation
holes

plastic cups
painted silver

spacers

thermocouple
junction

direction of
air flow

FIGURE 10. A simple shield for a temperature sensor in a growth chamber.

satisfactory. Resistance bulbs (wire wrapped around ceramic, for example, changing resistance with temperature) are a rather good compromise, having fairly good linearity of response without too high heat capacity (can be smaller than a postage stamp to perhaps pencil size).

Placement of the sensor is important, since there is almost always considerable temperature variation within a growth chamber or growth room. Ideally, the sensor should be placed among the leaves that are being studied. For many growth chamber applications, however, an investigator may be using the controlled environment facilities only to produce standard plant materials for research. In this case, more permanent placement of the sensor, perhaps to one side of the chamber, may be quite acceptable. Studies on temperature gradients should be carried out at the beginning of an experiment and possibly during it to provide a basis for choosing test plants from within the chamber.

As light intensities in growth chambers increase, the differences between leaf and air temperatures will also increase, as already shown. This may make it important to observe leaf temperature itself. The investigator might even want to control leaf temperature by sensing it and then using the information to adjust the cooling or heating system (Matsui and Eguchi, 1973). Thermocouples can be used to measure leaf temperature, although slight damage to the leaf and the stringing of wires with long-term installations may be minor problems. Small thermocouples can be used to briefly touch a leaf for single measurements (Tanner and Goltz, 1972; Pieters and Schurer, 1973). Surface temperature measurement by detection of infrared emitted from the leaf is an elegant way to approach the problem of leaf temperature. It is both remote and nondestructive. Most leaves are thin enough that one would not expect a significant temperature gradient within the leaf - although this is not true for succulent leaves or for stems. An infrared thermometer is an

ideal way to check on thermocouple measurements. But good
instruments are expensive and may require calibration (e.g.
determination of leaf emissivities) for accurate measurement.

CONTROLLING TEMPERATURE

Because the lights are the most significant source of heat
in most plant growth chambers, the problem of controlling
temperature is mostly one of heat removal. If the chamber is
being operated at temperatures below ambient, there will also
be conduction of heat into the chamber through the walls. During
the dark period and in other special circumstances, it may be
necessary to add heat to a growth chamber, but this can usually
be accomplished with simple electric heaters that respond to
the control system.

The most widely applied principle of temperature control
involves passing air over heat exchangers, the air temperature
changing by conductive and convective heat exchange. The coils
usually contain refrigerant that removes heat as it evaporates.
An alternative approach is to circulate a chilled fluid through
the heat exchanger. In either case, it is important to arrange
things so that the air first hits the coldest side of the heat
exchanger and exits the heat exchanger through the warmest side.

The problem of air flow in a growth chamber is important
in many respects. If air moves from one side to the other across
a chamber while the lights are on, it may be significantly
warmer on the exit side. Thus, many chambers are built so that
air enters from below the plants, passing up through the leaves.

Some investigators even advocate flow of air downward from
top to bottom so that air at the controlled temperature first
strikes the upper leaves which are receiving the most light.
Vertical air flow will in either case lead to vertical temperature gradients, but because of boundary layer effects on the
walls of the chamber, there will still be lateral gradients.

Of course, these can be reduced by increasing the rate of air
flow but at the peril of introducing too much mechanical
stimulation of the plants by wind. Some purists may also worry
about the appropriateness of moving air from beneath or above,
since this is such a rare situation in natural environments.
Yet no one can say exactly why upward or downward air flow might
be harmful, so most growth chambers do have air moving vertically
through the plants.

A good way to study the gradients in a chamber is to grow a
group of genetically identical plants (e.g., peas) in the chamber
and look for size and other differences. Flowering response of
cocklebur to a single inductive dark period, as well as growth,
was studied as a function of chamber position. Uniformity of
response was high (Salisbury, 1964).

An alternative approach to temperature control is to utilize
the principle of radiation instead of convection. Temperatures
of the walls and ceiling are controlled, and these exchange
radiant energy with the plants inside, influencing leaf temperature. Such chambers have been built (Mellor, Salisbury, and
Raschke, 1964), showing that the principle is feasible (Table 2).
It was extremely difficult to control the radiation environment
of the plant in a confined volume without strongly influencing
the air temperature by convective heat transfer between the
walls and the air. This was achieved by separating the walls
from the air in contact with the plants by a barrier of polyethylene film, which stopped convective heat transfer but
allowed transfer of thermal infrared.

It is nearly always desirable to have a capability for
separate day and night temperatures. This is easily achieved
with a time switch that changes from one controller to another.
Continuous change in temperature as a function of time has also
been offered by various companies. These systems have usually
been based upon the temperature set-point being changed by

TABLE 2. *Influence of the Thermal Radiation Environment upon Leaf Temperatures of Plants in the Dark.(From Mellor, Salisbury, and Raschke, 1964).*

Air temperature	Difference between leaf and air temperature at indicated ceiling temperatures		
	11.4°C	21.1°C	30.1°C
15°C	−1.5°C	−0.8°C	+1.7°C
25°C	−4.8°C	−0.9°C	+1.2°C
35°C	−5.1°C	−1.3°C	−0.9°C

24-hour rotation of a cam. These systems do not always function properly and have caused much trouble because it is difficult to make small adjustment during the course of an experiment. It is now possible to control temperature as a function of time through a microprocessor instead of a mechanical system. It hardly seems worthwhile, however, to have a capability for controlling temperature as a function of time if light intensity is not similarly controlled, and this can be done.

In relation to temperature control, it is appropriate to mention the transparent barrier often placed between the lamps and the plants. Air temperature around the lamps is controlled separately from air temperature around the plants because light output of fluorescent lamps is a function of their ambient temperature. The barrier reduces light input into the chamber by at least 10% under the best conditions, however, and for many applications such a plastic barrier may not be needed. It is certainly essential for studies conducted over wide temperature ranges (e.g., from close to freezing to 40°C).

Controlling humidity within a growth chamber will have some minor effect on leaf temperature, but this has usually been ignored (see Matsui and Eguchi, 1972 and 1973). As leaf temperature is recognized to be of increasing importance, the role of humidity should be studied more carefully.

In some applications, separate control of soil temperature
may be desirable. Soil temperatures in the field are almost
always higher during the day and lower during the night than they
are in a growth chamber. In the alpine tundra, soil temperatures
were virtually always lower than in a growth chamber where
alpine light conditions and air temperatures were simulated
(Spomer and Salisbury, 1968). It seemed possible that low soil
temperature might account for some of the plant growth responses
observed in the field but not in growth chambers (most notably,
induction of flowering in *Geum rossii*). Thus, the soil temperature controller shown in Figure 11 was constructed. Its principle
of operation is based on 4 vats of fluid (antifreeze plus water),
two of which are cooled by refrigeration coils and two of which
are heated by aquarium heaters. The fluid level is adjusted to
be in contact with the bottom of the soil containers, which are
about 30 cm deep. The soil in the containers above the chilled
fluid thus behaves much like soil above a permafrost. There is a
gradient of increasing temperature through the soil, and waves
of heat moving down, much as is observed in the field (Fig. 12).
The gradients are reversed at night in the soil containers above
the heated fluid. Other soil temperature controllers have been
constructed (e.g., Spomer, 1976), sometimes by placing heating
and cooling coils directly in the soil. With coils in the soil,
there will still be temperature gradients, but they can be
minimized if a sufficient number of tubes is used.

TEMPERATURE CONTROL SYSTEMS

Two somewhat different approaches can be used in controlling
temperature: an on-off system or a modulated one. Both are
negative feedback devices. The on-off system is by far the
most common. A valve that allows refrigerant to flow into a
cooling coil is opened or closed in response to the thermostat.

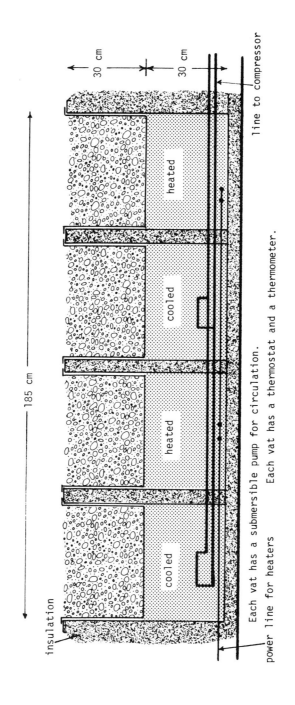

FIGURE 11. Design of a unit for controlling soil temperatures by controlling temperatures of fluids in deep vats below the soil vats. The unit was constructed at Utah State University.

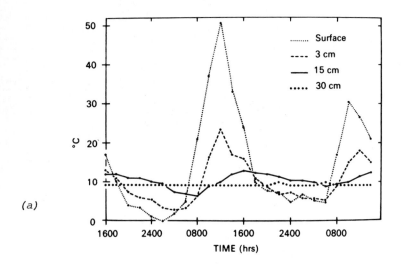

(a)

FIGURE 12. *(a) Soil temperatures at various depths as a function of time in the alpine tundra of Rocky Mountain National Park in Colorado (from Salisbury, et al., 1968) and (b) in the soil vats illustrated in Figure 11 (data courtesy of Raymond Wheeler). Soil temperatures in the field remained steady at 30 cm depth, as in the vats, where temperatures are controlled at that depth. Note the wave of heat moving down in the soil vat above cold fluid, much as in the alpine tundra (and in spite of considerable difference in the patterns of air temperatures above the soil). The wave of heat also moves down above the heated vat, but the gradient is inverted during the night.*

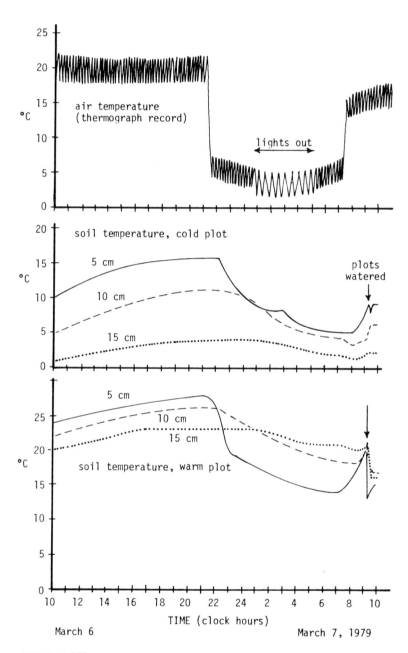

FIGURE 12b

A small system at Utah State University stabilizes the temperature in a room by turning the flow of cold water on or off through a heat exchanger.

In an on-off system there is always a considerable fluctuation in temperature. If the cooling coils have the capability of producing very low temperatures, then when they are suddenly turned on there will be a rapid cooling of the air but probably a lag in response of the control system sensor. Thus the air temperature will drop well below the set point before the sensor has time to respond and turn off the system. There is a systematic and sometimes extreme temperature fluctuation (see air temperature record in Figure 12 for an extreme example). If the sensitivity of the sensor is increased, then the system cycles on and off too frequently. This problem is usually solved by carefully matching the capacity of the cooling system to the needs of the investigator. Additional cooling systems could be added when especially low temperatures are desirable.

In a modulated system, the *rate* of cooling or heating is a function of the deviation of detected temperature from the set point - the greater the deviation the more rapid the cooling or heating that will reestablish the temperature. Ideally, some rate of cooling (or heating) might be achieved that would just balance the input of heat (or cold) so that temperature might remain steady (Fig. 13). Either an on-off or a modulating system must have a neutral range where neither cooling nor heating is taking place, and the modulated system should be able to maintain a highly steady temperature, perhaps just outside this range. Some systems are modulated by controlling the rate at which refrigerant enters the cooling coils or by controlling the temperature of the entering refrigerant. In a system built at Colorado State University many years ago (Salisbury, 1964), the rate of flow of chilled fluid through the cooling coils was controlled by a Johnson pneumatic control system (Fig. 14).

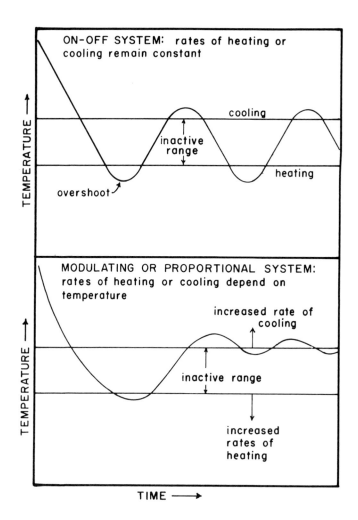

FIGURE 13. Expected temperature patterns in on-off and modulated control systems. (From Salisbury and Ross, 1969.)

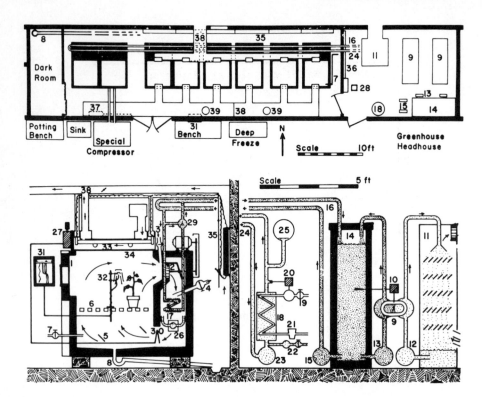

FIGURE 14. Outline of a system of growth chambers built at Colorado State University and utilizing a modulated temperature control by controlling the rate of flow of chilled fluid through heat exchangers in the chambers. Numbers refer to a table that will not be reproduced here. (From Salisbury, 1964.)

A modulating temperature system should provide a more suitable temperature, especially in a small chamber. In a large growth room with a large volume of air, many plants, large wall space, etc., there is a certain buffering, thermal capacity against the rapid temperature changes brought about by an on-off system, but in a small chamber, this buffering capacity is minimal. Furthermore, low cooling rates extended over a greater portion of the time provide warmer surfaces on the cooling coils, which reduce condensation of atmospheric moisture and the sharp reduction in humidity during cooling that is often characteristic of an on-off system. A system with chilled fluid (e.g., ethylene glycol and

water) is desirable, because the fluid temperature can be fairly
warm. It may not be practical to use fluid at a temperature
above the freezing point, however, since every exposed valve or
pipe condenses moisture and drips. With fluid a few degrees
below freezing, solid ice forms on exposed pipes and valves and
does not drip. To some, at least, ice seems preferable to
puddles.

Having presented what seems to be a rather convincing
argument for a modulating control system, it must be added that
experience has not demonstrated a great superiority for such
systems. Temperature in the Colorado system fluctuated almost
like it does in any on-off system. Representatives of the growth
chamber industry also report that modulating systems offer few
practical advantages.

DESIRED TEMPERATURE RANGES

The desired temperature capabilities of growth chambers will
depend upon whether the investigator is studying plants of the
Antarctic, those from Death Valley, or Michigan sugar beets.
Obviously, any temperature range that might be used in plant
studies can be provided with available engineering.

A much more important problem concerns the effects of
temperature fluctuations that arise as a natural function of a
control system. How do plants respond to these fluctuations if
they respond at all? One study addressed this problem (Evans,
1963). Short term fluctuations of about 2.5°C every 2 minutes
resulted in significantly greater leaf area and dry weight than
with fluctuations of only 0.5°C every 2 minutes. In short, these
highly preliminary results published over a decade and a half
ago suggest that the normal temperature fluctuations in the
growth chamber may promote plant growth. Their counterpart would
be the "climatic noise" observed in the field, although field
temperature fluctuations are greater and more random than those

in a growth chamber (Fig. 15). It has seemed appropriate for a
long time that studies should be made of these matters. What
would be the optimum periods and amplitudes of fluctuation? Is
a "square wave" fluctuation between day and night equivalent to a
more sinusoidal change? Evans, 1963, states that he also tested
this but does not report his results clearly.

SOME PROPOSALS

There are some things related to temperature control for plant
growth that have seemed in need of testing for many years - or
that have been tested but not reported. Now is a chance to get
some of these ideas into print! (No doubt, many investigators
have thought of these same things; at least one has been tested
in the Wisconsin Biotron.)

Minimum Fluctuation Chamber

In studies of energy transfer between a leaf and its environment, it became highly desirable to reduce air temperature
fluctuations to only a small fraction of a degree (Drake and
Salisbury, 1972). This would also be desirable as a control in
studies of plant responses to temperature fluctuations. An
approach to this problem is to pass the air through a large heat
exchange coil through which fluid is being circulated at almost
the exact temperature needed in the chamber. It is relatively
easy to control temperature of a large volume of fluid to within
0.01 to 0.1°C. So that this last heat exchange coil only does a
little of the required cooling (or heating) and thus itself
remains highly stable, the air might first pass through one or
more other heat exchangers, these also containing chilled fluids,
but probably at a temperature that deviated more from the desired
temperature than the temperature of the fluid in the final coil
(Fig. 16).

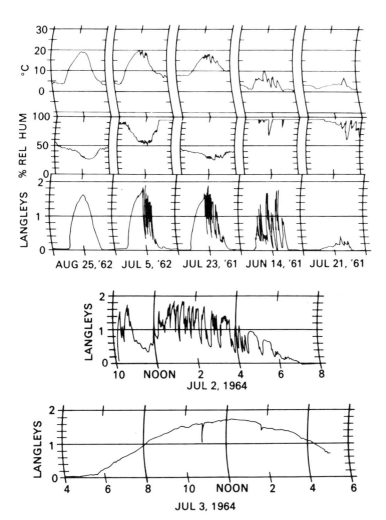

FIGURE 15. Some patterns of temperature, relative humidity, and light intensity measured in the alpine tundra of Rocky Mountain National Park, Colorado. In the upper figure, representative days are arranged from a clear day with minimal fluctuations on the left through partially cloudy days with high variations in environmental factors to a cold and cloudy day that is relatively stable. The lower part of the figure shows light intensity on two days on Independence Pass in Colorado, one partially cloudy (high variability), the other almost clear. (From Salisbury, et al., 1968.)

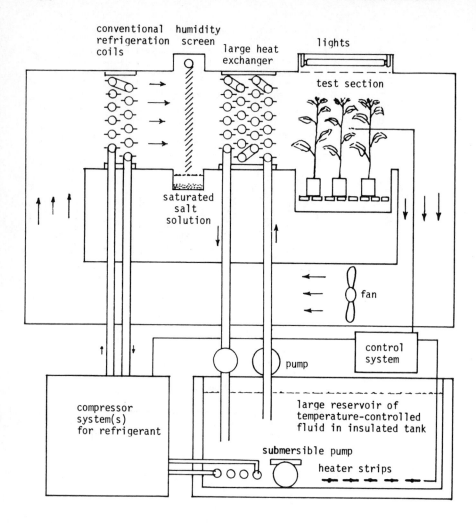

FIGURE 16. A concept for a chamber with minimal temperature fluctuations and controlled humidity.

In the heat-transfer studies, instead of a series of coils, a sheet-metal wind tunnel within which air was continually recirculated was placed in a room in which temperature was already controlled (although with considerably fluctuation), and the air was passed through a large heat exchange coil containing fluid of

the desired temperature just before the air crossed the leaves
being studied. Extremely stable air temperatures were produced,
as measured with fine wire thermocouples.

The same study utilized a system for humidity control that is
not often used. Relative humidity in a closed space above
saturated salt solutions is quite constant over a fairly wide
temperature range. By passing the air in the tunnel through a
curtain of saturated salt solution (calcium chloride and also
ethylene glycol solutions were used), humidity could be fairly
well stabilized. This system was not as effective as the
temperature control system, however, although the principle was
clearly demonstrated.

Studying the Role of Fluctuating Temperatures

It would be good to modify a steady state system such as that
just described so that both amplitude and period of rhythmical
temperature swings could be controlled. Another variation of
such a system would be to vary temperature according to some
pattern measured in a field situation. By duplicating temperature, radiation, humidity, and wind fluctuations in a chamber,
would it be possible to exactly duplicate growth of plants in
the field?

One problem is how to change temperatures in a chamber
rapidly and predictably. Connor (1966) worked out such a system
on paper. The principle was to have two or more separate
temperature control systems (sets of heat exchangers) and then to
direct the flow of air with a swinging gate first through one of
these systems and then through another as temperature fluctuations
were specified by the control system. Such an arrangement should
have the capability for rapid (less than a minute) and extensive
(several degrees) temperature fluctuations. It could be
controlled by a microprocessor. Connor's system suggested control
with punched or magnetic tapes, so that temperature fluctuations

in a field situation could be readily duplicated. Secondary
computer tapes could be produced to reduce (or amplify) the
fluctuations to any degree that seemed desirable, thereby making
it possible to study plant responses (if any) to the fluctuations.

Such a system is being used in air conditioning large
buildings (e.g., the Logan, Utah, Latter-day Saint Temple). Both
warm (hot in winter) and cool (cold in summer) air is continually
circulated through the building's system. Thermostats in each
room modulate the mixture of heated or cooled air entering that
room, providing stability within about 0.25°C all year.

Gradient Chambers

The Climatron in St. Louis, designed by Went (1963), attempted
to produce two cross gradients; one in temperature and the other
in humidity. That is, air that entered on one side gradually
warmed as it moved through the structure, becoming several degrees
warmer on the opposite side (a situation common to most greenhouses, of course). One side of the structure had high humidity,
and the other side lower humidity. It was hoped that these two
cross gradients would simulate natural environments from dry,
cool areas to hot, moist jungles with appropriate species growing
in the proper places. (Went appeared satisfied with his system,
but the concept has now been abandoned in the St. Louis
Climatron.)

Such an approach could also be used on a smaller scale in a
fairly large growth chamber. A temperature gradient seems
logical, and if artificial light is used, a cross gradient in
light intensity would be appropriate. Growth and response of
individual plants from a single chamber could be plotted on a
3 dimensional graph, showing growth as a function of both
temperature and light.

The complication is that changing temperatures leads to changing humidities, thereby somewhat confusing the results. This problem proved insurmountable with such a gradient chamber in the Wisconsin Biotron (Tibbitts and Kozlowski, personal communication). Perhaps one might cool the air as it goes along, rather than allowing it to heat in response to the radiant energy input. The air might pass through a series of baffles, each containing a heat exchanger. Humidity would remain high (close to 100%) in all the system. Another approach might be to increase humidity along the system as the air was allowed to warm, perhaps by adding steam. The addition of water as mist would cool the air somewhat as the water evaporated. It might be worthwhile to build a cross gradient (temperature - light) chamber and simply ignore the humidity problem until a little experience was obtained. Under some circumstances (dense plant cover, thin air layer), transpiration might maintain relatively constant humidity as the air warmed. The alternative to obtaining data on optimum temperature and light conditions for plant production is a number of chambers and/or many duplicated runs.

It should be evident from this discussion that, while controlling temperatures in artificial plant environments continues to have challenging problems, most of the temperature control required for adequate plant growth can readily be achieved with current levels of technology - is indeed being achieved. But the challenges will increase as more intense light sources are used.

REFERENCES

Connor, G. I. (1966). Design of a Wind Tunnel Growth Chamber. M.S. Thesis, Colorado State University, Fort Collins, Colorado.

Drake, B. G., Raschke, K., and Salisbury, F. B. (1970). Temperatures and transpiration resistances of *Xanthium* leaves as affected by air temperature, humidity, and wind speed. *Plant Physiol.* *46*, 324-330.

Drake, B. G., and Salisbury, F. B. (1972). Aftereffects of low and high temperature pretreatment on leaf resistance, transpiration, and leaf temperature in *Xanthium*. *Plant Physiol.* *50*, 572-575.

Evans, L. T. (1963). Extrapolation from controlled environments to the field. *In* "Environmental Control of Plant Growth", (L. T. Evans, ed.), pp. 421-437. Academic Press, New York and London.

Gates, D. M. (1965). Heat transfer in plants. *Sci. Amer. 213*, 132-142.

Gates, D. M., and Schmerl, R. B., eds. (1975). "Perspectives in Biophysical Ecology." Springer-Verlag, New York.

Jaffe, M. J. (1973). Thigmomorphogenesis: The response of plant growth and development to mechanical stimulation. With special reference to *Bryonia dioica*. *Planta 114*, 143-157.

Jaffe, M. J. (1976). Thigmomorphogenesis: A detailed characterization of the response of beans *(Phaseolus vulgaris L.)* to mechanical stimulation. *Z. Pflanzenphysiol. 77*, 437-453.

Lange, O. L., Losch, R., Schulze, E. D., and Kappen, L. (1971). Responses of stomata to changes in humidity. *Planta 100*, 76-86.

Matsui, T., and Eguchi, H. (1972). Effects of environmental factors on leaf temperature in a temperature controlled room. II. Effect of air movement. *Environ. Control in Biol. 10*, 105-108.

Matsui, T., and Eguchi, H. (1973). Feedback control of leaf temperature. *Environ. Control in Biol. 11*, 55-63.

Meidner, H., and Mansfield, T. A. (1968). "Physiology of Stomata." McGraw-Hill, New York.

Mellor, R. S., Salisbury, F. B., and Raschke, K. (1964). Leaf
    temperatures in controlled environments. *Planta 61*, 56-72.

Mitchell, C. (1977). Influence of mechanical stress on auxin-
    stimulated growth of excised pea stem sections. *Physiol.
    Plant 41*, 129-134.

Neel, P. L., and Harris, R. W. (1971). Motion-induced inhibition
    of elongation and induction of dormancy in *Liquidambar*.
    Science 173, 58-59.

Pieters, G. A., and Schurer, K. (1973). Leaf temperature measure-
    ment I. Thermocouples. *Acta Bot. Neerl. 22*, 569-580.

Raschke, K. (1960). Heat transfer between the plant and the
    environment. *Annu. Rev. Plant Physiol. 11*, 111-126.

Salisbury, F. B. (1963). "The Flowering Process", p. 161.
    Pergamon, Oxford, London, New York, Paris.

Salisbury, F. B. (1964). A special-purpose controlled-environment
    unit. *Bot. Gaz. 125*, 237-241.

Salisbury, F. B., and Ross, C. W. (1969, 1978). "Plant Physiolo-
    gy," First and second editions. Wadsworth Pub. Co., Belmont,
    California. (Note: the discussion of heat transfer is
    somewhat more extensive in the first edition.)

Salisbury, F. B., and Spomer, G. G. (1964). Leaf temperatures
    of alpine plants in the field. *Planta 60*, 497-505.

Salisbury, F. B., Spomer, G. G., Sobral, M., and Ward, R. T.
    (1968). Analysis of an alpine environment. *Bot. Gaz. 129*,
    16-32.

Spomer, G. G. (1976). Simulation of alpine soil temperature
    conditions. *Arctic Alpine Res. 8*, 251-254.

Spomer, G. G., and Salisbury, F. B. (1968). Eco-physiology of
    Geum turbinatum and implications concerning alpine environ-
    ments. *Bot. Gaz. 129*, 33-49.

Tanner, C. B., and Goltz, S. M. (1972). Excessively high
    temperatures of seed onion umbels. *J. Amer. Soc. Hort. Sci.
    97*, 5-9.

Van Wylen, G. J., and Sonntag, R. E. (1976). "Fundamentals of Classical Thermodynamics", 2nd Edition. Wiley, New York.

Victor, T. S., and Vanderhoef, L. N. (1975). Mechanical inhibition of hypocotyl elongation induces radial enlargement. *Plant Physiol.* *56,* 845–846.

Went, F., and the Editors of Life (1963). "The Plants", p. 131. Life Nature Library, Time Incorporated, New York.

Wheeler, R. M., and Salisbury, F. B. (1979). Water spray as a convenient means of imparting mechanical stimulation to plants. *Hortscience 14,* 270–271.

TEMPERATURE: CRITIQUE I

C. B. Tanner

Department of Soil Science
University of Wisconsin
Madison, Wisconsin

Some features of air temperature measurement will be discussed. Then an aspect of infrared thermometry of leaves will
be considered, followed by a review of direct measurement of
leaf temperature.

AIR TEMPERATURE MEASUREMENT

Thermometers used for air temperature measurement usually
are subject to a net exchange of radiation, such as from growth
chamber lamps or from cold walls as described by Salisbury. The
sensor then either heats or cools radiatively until the convective exchange with the air equals the net radiation exchange.
The radiation error, the difference between the thermometer
temperature and air temperature, is

$$\Delta T = R_n/h \tag{1}$$

where R_n is the net radiation flux density and h is the convective heat transfer coefficient. The radiation error can be
decreased by shielding the sensor from radiation. However,
radiation shields may either heat or cool and then exchange heat
with the thermometer either convectively or radiatively;

117

unaspirated shields should always be tested. Increasing the convective transfer coefficient also will decrease the radiation
error. The transfer coefficient can be increased either by
increasing the airflow past the sensor or by making the sensor
smaller. This can be illustrated with small cylindrical sensors
such as thermocouples or wire resistance thermometers. Similar
principles apply to thermistors and other thermometers, but very
samll thermocouples and resistance wire thermometers can be
readily fabricated in small sizes and are discussed here. The
transfer coefficient, h, for wires under transverse flow can be
estimated from

$$h\ D/k = A + B(V\ D/\nu)^{1/2} \tag{2}$$

where D is the wire diameter, V is the flow velocity normal to
wire, k and ν are the thermal conductivity and kinematic viscosity
of the air, and A and B are dimensionless constants. Grant and
Kronauer (1962) give values of A for different wire length/
diameter ratios as in Table 1. B is a function of the Prandtl
number of the fluid, and is 0.51 for air. Collis and Williams
(1959) give formulas for free convection from fine wires (no
forced flow). Using the Collis and Williams expression for free
convection and equation (2), we can calculate the approximate
temperature rise for thermocouples of different diameters for a
mean radiation (<4 μm) flux density to the wire of 300 w m^{-2}
(high for growth cabinets), an absorptance of 0.5 at wavelengths

TABLE 1. *Values of A for Different Length/Diameter Ratios
(From Grant and Kronauer, 1962).*

ℓ/D:	1	3.72	10	33	100	320	1000
A:	2.2	1.0	0.66	0.47	0.38	0.29	0.25

<4 µm (reasonable for thermocouple alloys) and zero thermal radiation (>4 µm) emissivity (0.1 to 0.3 is reasonable). Estimates for these severe conditions are given in Table 2, mainly to illustrate the effect of wire size.

It is clear that with usual chamber ventilations, thermocouples $\leq$75 µm can be used without shielding under fairly high radiation if <0.5°C error can be tolerated; however, using couples as large as 250 µm (30 ga B&S) may lead to significant error. Unaspirated radiation shield arrangements used with larger thermometers should be tested using 25-µm thermocouples, or finer, to assure they are performing adequately.

It is noted that if flow is axial to the leads, the transfer coefficient decreases below that for transverse flow. Wesely, Thurtell, and Tanner (1970) give construction details of a 25-µm wire resistance thermometer that can be adapted for environmental chamber use and which has negligible radiation error. The time-constant of fine wires is short and electronic integration of signals usually is helpful to minimize fluctuations in output.

LEAF TEMPERATURE MEASUREMENT

As discussed by Salisbury, the temperature of the leaf is a result of its energy balance (radiative, convective, and latent heat exchange); the leaf may be cooler or warmer than the air.

TABLE 2. *Estimated Temperature Rise ($°K$) for Four Wire Sizes and for the Conditions Given in the Text*

Air velocity (cm s^{-1})	Wire diameter (µm)			
	25	75	125	250
0	0.5	1.2	1.7	2.6
10	0.25	0.6	0.8	1.3
40	0.2	0.4	0.5	0.8
100	0.1	0.3	0.4	0.6

If the thermometer blocks radiation, alters air circulation, or affects transpiration, the temperature of a thin leaf will alter in a few seconds (Wiegand and Swanson, 1973). It is important not to affect the leaf energy balance appreciably during measurement unless the measurement can be made in 1 to 3 seconds. Because leaves have relatively low thermal conductivity and heat capacity per unit area, thermometers with low thermal capacity are necessary. The temperature difference between leaf and bulk air exists across a thin air layer next to the leaf of the order of 1 mm in thickness. Thus, conduction of heat through the thermometer across this thin boundary layer can cause appreciable error, particularly if the leaf is thin (low heat capacity per unit area) and thermal contact between leaf and thermometer is poor. Because of the above problems, infrared thermometry, small thermocouples, and small thermistors have been used. Data obtained with fine-wire thermocouples, to be discussed later, indicate serious errors easily could arise with thermistors which are usually thicker than fine-wire thermocouples, and make poor contact with a leaf. Accordingly, discussion here is restricted to radiation thermometers and fine-wire thermocouples.

Infrared Thermometers

At environmental temperatures, nearly all of the emitted (thermal) radiation is in the 4- to 100 μm band. Many radiation thermometers include band-pass filters in the atmospheric window (approximately 8 to 13 μm). This filtering is not necessary for use in growth chambers and wider band response (>4 μm) is not only acceptable but may prove helpful in increasing sensitivity. In the spectral range >4 μm, leaves are highly absorbing because of the liquid water contained and since reflectance of most plants varies little with wavelength (Gates and Tantraporn, 1952), nearly uniform spectral emissivity can be assumed; measurements indicate leaf emissivity is near 0.95. The thermal radiation

flux density, R_T, to which the thermometer responds, is made up
of emitted radiation and reflected background thermal radiation,
B, from ceiling, walls, etc. Although the spectral density of
emitted and reflected radiation must be considered and weighted
by the spectral reflectance of leaf and thermometer spectral
response (Fuchs and Tanner, 1966; 1968), the error resulting from
leaf emissivities less than unity can be illustrated using simple,
spectrally-integrated terms. In simplest form, the infrared
thermometer "sees"

$$R_T = \varepsilon_L \sigma T_L^4 + (1-\varepsilon_L)B \tag{3}$$

where T_L is Kelvin leaf temperature, ε_L is leaf emissivity; σ is
the Stefan-Boltzmann constant; and B is the background radiation.
The apparent Kelvin leaf temperature is found as $T_a = (R_T/\sigma)^{0.25}$
If $B = \sigma T_L^4$ (effective temperature of the background is the same
as T_L, as would be approximated during the dark period), $R_T = \sigma T_L^4$ and the apparent temperature equals T_L. However, during
the light period, the temperature of the hot lamps is much greater
than T_L and chamber walls possibly are at quite different temperatures. Under this condition, corrections are needed, and may
be large (>2°C) and are difficult to make. However, if the
radiation thermometer has fast response, has a narrow viewing
angle, and if it is fitted with a cone with a low-emissivity
interior such as machined aluminum (Fig. 1), then the leaf temperature can be measured directly by placing the aperature of
the cone against a leaf. The effective emissivity, $\varepsilon_e = R_T/T_L^4$, when viewed through the aperature at the cone apex, can
be approximated as:

$$\varepsilon_e = \frac{\varepsilon_L + (1-\varepsilon_L)\varepsilon_c(1-f)[1-4(T_m^3/T_L^4)(T_L - T_c)]}{1-(1-\varepsilon_L)(1-\varepsilon_c)(1-f)} \tag{4}$$

where ε_c is the effective emissivity of the cone as viewed by
the leaf; f is (area of view port)/(area of cone walls, excluding

leaf area); T_c is Kelvin cone temperature; and $T_m = (T_c + T_L)/2$.
The approximation is valid for small f.

Sparrow and Cess (1966; p. 168) show that with wall emissivity
of 0.1 (high for aluminum), $\varepsilon_c \simeq 0.3$ for a 60° total included
apex angle and 0.18 for 120° included angle; ε_c will decrease
about two-fold for clean aluminum with emissivity of 0.05.
Assuming ε_L is 0.9 (low), ε_c = 0.15, T_L = 300°K, T_c = 290°K, and
f = 0.1, we find ε_e = 0.987. The error resulting from assuming
ε_e = 1 is about -1.0°C. Since ε_L is typically about 0.95, a
similar calculation gives ε_e = 0.994, and the error resulting
from assuming ε_e = 1 is about -0.5°C. Thus, major uncertainties
from B and ε_L are avoided, and relatively minor and simple corrections of a fairly well-defined system can be made.

The leaf will change temperature in seconds when the cavity
is placed on the leaf so it is advisable to record the thermometer
output. The skewed cut at the end of the cone is helpful, when
specularly reflecting surfaces are viewed, to avoid back reflection of the radiation beam emitted by the radiometer. The same
arrangement is helpful in calibrating the radiometer by placing
it against a metal block of known temperature, which has been
coated with high emissivity material; heavily anodized aluminum
is a useful reference.

Thermocouples

Many thermocouple and thermistor methods for measuring the
temperature of leaves or floral parts have been reported. Lange
(1965) reviews many earlier leaf temperature measurements. A
few later examples include Idle (1968), Gale, Manes, and Poljakoff-
Mayber (1970), Linacre and Harris (1970), Beadle, Stevenson, and
Thurtell, (1973), and Pieters and Schurer (1973). This is not an
exhaustive listing, and other references may be found in the
articles cited. The concern here mainly is with making better
use of leaf thermocouples.

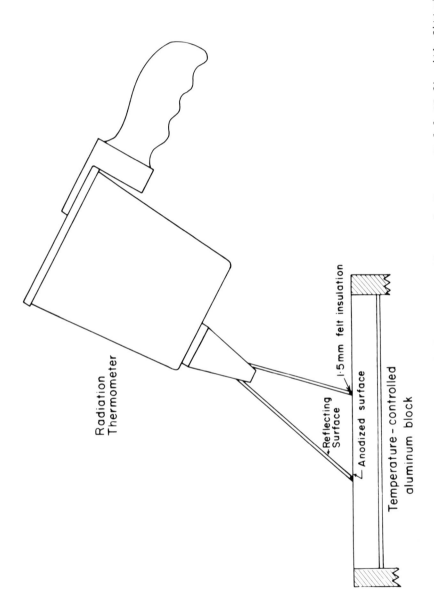

FIGURE 1. Schematic diagram of infrared thermometer (focused, Barnes Model IT-3) with fitted conical chamber.

Thermocouples usually are placed against the most shaded leaf surface in order to minimize radiation errors. Butt-welded thermocouples have been threaded through veins and lamina for support, they have been either taped or cemented to the leaves, and they have been held in contact with leaves by positioning the lead wires. Devices also have been constructed to clamp thermocouples and thermistors against leaves (Lange, 1965; Gale, Manes, and Poljakoff-Mayber, 1970 ; Linacre and Harris, 1970). Linacre and Harris held a 0.5-mm thermistor cemented to a 1.2-mm ceramic rod against the leaf surface; on the basis of tests by others with fine-wire thermocouples, this large thermometer would give large errors.

Idle (1968) illustrated a convenient set-up for testing contact thermocouples, by placing them against a bare copper surface of known temperature under fan ventilation. With unspecified flow, he tested 120-μm copper/constantan couples and showed with only the junction in contact with the copper plate, the measured temperature difference to air was only 40% of the true difference (60% error). With 3 mm of lead (25 wire diameters) contacting the plate, a 30% error obtained and with a small film of oil aiding contact with the 3 mm of lead, the error was only 5%. The error should decrease with lower ventilation (thicker boundary layer) and with lower conductance wires. Gale, Manes, and Poljakoff-Mayber (1970) used a 100 μm stainless steel/50-μm constantan couple tensioned against the leaf with a clamp. The poorly-designed clamp interfered greatly with the leaf energy balance so that measurement within a "few seconds" was required. Comparisons with a 50-μm iron/constantan couple threaded in the mesophyll of a chard leaf indicate accuracies of $\pm 0.5^{\circ}$C obtained with leaf-air gradients of 2 to 3°C. Agreement with an infrared thermometer was within the accuracy of the infrared thermometer; however, their discussion does not indicate that needed corrections of the infrared thermometer were considered. They do show that holding a couple to leaves with tape can cause considerable error, and that under solar radiation loading, couples threaded in large veins were at significantly higher temperatures than the rest of the leaf.

The most definitive investigations of possible thermocouple error are those of Beadle, Stevenson, and Thurtell (1973) and Pieters and Schurer (1973). Beadle, Stevenson, and Thurtell compared 25-μm Evanohm/constantan[1] resistance-welded junctions with commercial copper/constantan junctions made from 125-μm wire but with beaded junctions about 300-μm diameter. Both types of junctions were crimped to large thermocouple lead wire and the junction and some fine-wire lead was tensioned against the shaded side of a leaf in a cuvette by springing the larger lead wire. With leaf boundary layer transfer coefficients of about 20 cm s^{-1}, the commercial junction measured about 0.5 the air-to-leaf temperature difference that was measured by the 25-μm junction. The difference was due both to poor contact between the large bead and leaf and to heat conduction mainly through the thin (order of 100 μm) boundary layer via the fine copper leads and large bead.

Pieters and Schurer (1973) irradiated on one side, a black-painted, 3-mm thick, copper plate placed in a cuvette at 25 cm s^{-1} flow. The plate temperature was measured with thermocouples in the plate and compared with surface temperature measurements made by holding thermocouples to the shaded surface of the plate. Three types of thermocouples were used: soldered, 100-μm copper/constantan and butt-welded chromel/constantan using both 100- and 25-μm wire (the butt-welded junctions were formed in a 5-mm circular loop with the junction opposite the leads). The data are difficult to interpret because the thermal resistance of the paint could be appreciable and because it was not clear whether a length of lead and junction contacted the plate or if contact

---

[1]*Evanohm is a material made by Wilbur Driver Co., Newark, New Jersey, for wire-wound resistors. It has low thermoelectric coefficient against copper ($\simeq$ +0.2 v/°C) and can serve in place of copper.*

was made only by the junction. The $(T_L - T_a)$ difference measured
with 100-μm couples was about 70% of the actual difference.
Couple materials were not given, but likely were chromel/constan-
tan. The authors indicate that with 25-μm couples pressed against
the surface, an error of less than 1°C in a 15°C difference
resulted. Other data indicated an error of 1°C in a 5°C
difference could be expected when the couple was not forced
against the surface. Since Pieters and Schurer mention using
"2 gm of pressure" on the 25-μm couple, it is possible that they
were forcing the couple into the paint and decreasing some of the
thermal resistance of the paint.

It is clear from the data in the literature that accurate
leaf temperature measurement requires good contact between leaf
and junction, including a length of lead wire contacting the
leaf. Beaded couples increase contact resistance and should be
avoided. Additionally, heat flow through the leads should be
minimized by choice of materials and using small wires. The
thermal conductivity of some thermocouple materials is given in
Table 3, along with the relative heat-flow as compared to copper
thermocouple wire. Heat flow through copper leads is 20-fold
that through other materials. Moreover, the other materials
are 3- to 6-fold stronger and resistance weld readily. Copper
has only disadvantages for any leaf thermocouple. The relative
heat flux of the smaller wires in Table 3 is an overestimate

TABLE 3. *Thermal Conductivities (Watt $cm^{-1}°C^{-1}$) of Thermocouple Materials and Relative Heat Flux for Different Wire Size
as Compared with 125 μm Copper*

	Copper	Evanohm[1]	Chromel-P	Constantan
Thermal conductivity	3.9	0.16	0.20	0.22
Relative heat flux				
125 μm	100	4.10	5.13	5.64
75 μm	36	1.48	1.85	2.03
25 μm	4	0.16	0.20	0.22

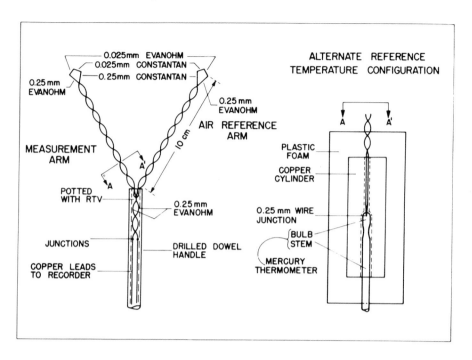

FIGURE 2. Configuration of thermocouple thermometer for leaf to air temperature difference (left) and leaf temperature (right). (From Tanner and Goltz, 1972).

since the heat transfer coefficient increases as size decreases and they would adjust better to the temperature gradient in the boundary layer.

We have used the arrangement shown schematically in Fig. 2 (Tanner and Goltz, 1972). The 25-μm junction is resistance welded to 250-μm supporting wires. The 25-μm wires can be welded at an angle to the 250-μm wires so that when the support wires are placed against a leaf, the junction and lead wires are tensioned against the leaf. Tests on an electrically-heated copper plate under fan ventilation between >50 cm s^{-1} (arrangement similar to that of Idle, 1968) provided air to plate gradients within 5% of those measured with a fine constantan junction soldered to the plate.

The most important question is how accurately the leaf temperature must be determined. The spatial variation over a leaf
may be more than 1°C and chamber air temperature may fluctuate
similar amounts. Except for special investigations with cuvettes,
studies under large temperature gradients (high radiation load;
low ventilation) and studies on transpiration and spatial variation of leaf temperature, 75-μm couples should suffice; these are
much more convenient to fabricate and use. If leaf damage is
inconsequential, dipping the junction in oil prior to contact
will help (Idle, 1968).

Fabrication of Thermocouples

The Ewald[1] resistance welders, suggested by Beadle, Stevenson,
and Thurtell (1973) are satisfactory for 25-μm to 250-μm wire.
They can be used on fine resistance thermometers as well as thermocouples. Their tweezer hand piece is useful for welding small
wires to larger wires. Wires are held on a card for welding by
small pieces of tape, across a punched hole in the card. The
weld should imbed wires into each other to give a junction thickness about equal to the wire diameter. Free ends can be cut away
with cuticle scissors. A short length of free wire beyond the
junction can be left to aid heat transfer between the wire and the
leaf. With 75-μm wire, or larger, the junction can be held in
tweezer and free ends broken off by flexing, if wished. We have
broken free ends away from a 75-μm Evanohm/constantan junction
and pulled the leads in opposite directions to provide couples
with the appearance of butt welds.

Evanohm requires special solders, so that either crimped or
welded junctions are more convenient. Crimped junctions between
Evanohm and copper are most convenient. We have found small

[1]*Ewald Instrument Corporation, Kent, Connecticut.*

copper sleeves[1] to be inexpensive and reliable for crimp connections. Crimp connections with these sleeves provide a more
reliable and convenient junction with all thermocouple materials
than soldering, provided the large joint size is permissible.
Resistance welding is most convenient for connections between
250-μm wires and smaller.

REFERENCES

Beadle, C. L., Stevenson, K. R., and Thurtell, G. W. (1973).
    Leaf temperature measurement and control in a gas-exchange
    cuvette. *Can. J. Plant Sci. 53,* 407-412.
Collis, D. C., and Williams, M. J. (1959). Two-dimensional
    convection from heated wires at low Reynolds numbers. *J.
    Fluid Mech. 6,* 357-384.
Fuchs, M., and Tanner, C. B. (1966). Infrared thermometry of
    vegetation. *Agron. J. 58,* 597-601.
Fuchs, M., and Tanner, C. B. (1968). Surface temperature mea-
    surements of bare soils. *J. Appl. Meteorol. 7,* 303-305.
Gale, J., Manes, A., and Poljakoff-Mayber, A. (1970). A rapidly
    equilibrating thermocouple contact thermometer for measurement
    of leaf-surface temperatures. *Ecology 51,* 521-525.
Gates, D. M., and Tantraporn, W. (1952). The reflectivity of
    deciduous trees and herbaceous plants in the infrared to 25
    microns. *Science 115,* 613-616.
Grant, H. P., and Kronauer, R. E. (1962). Fundamentals of hot
    wire anemometry, *In* "Symposium on Measurement in Unsteady Flow",
    pp. 44-53. Amer. Soc. Mech. Engr., 345 E. 47th Street, New
    York.

[1]
*No. P3A copper sleeves, 0.044-inch I.D. x 0.08-inch O.D. x
0.375-inch long. Berkley and Co., Spirit Lake, Iowa; similar
nickel sleeves, suited to high temperature use, also are available.*

Idle, D. B. (1968). The measurement of apparent surface tempera-
    ture. *In* "The Measurement of Environmental Factors in Terres-
    trial Ecology" (R. M. Wadsworth, ed.), pp. 47-57. Blackwell
    Scientific Publs., Oxford.
Lange, O. L. (1965). Leaf temperatures and methods of measure-
    ment. *Arid Zone Res. 25,* 203-209.
Linacre, E. L., and Harris, W. J. (1970). A thermistor leaf
    thermometer. *Plant Physiol. 46,* 190-193.
Pieters, G. A., and Schurer, K. (1973). Leaf temperature mea-
    surement. I. Thermocouples. *Acta Bot. Neerl. 22,* 569-580.
Sparrow, E. M., and Cess, R. D. (1966). "Radiation Heat Trans-
    fer." Brooks/Cole Publ. Co., Belmont, California.
Tanner, C. B., and Goltz, S. M. (1972). Excessively high
    temperatures of seed onion umbels. *J. Amer. Soc. Hort. Sci.
    97,* 5-9.
Wesely, M. L., Thurtell, G. W., and Tanner, C. B. (1970). Eddy
    correlation measurements of sensible heat flux near the earth's
    surface. *J. Appl. Meteorol. 9,* 45-50.
Wiegand, C. L., and Swanson, W. A. (1973). Time constants for
    thermal equilibration of leaf, canopy, and soil surfaces with
    change in insulation. *Agron. J. 65,* 722-724.

TEMPERATURE: CRITIQUE II

R. P. Searls

Sherer Environmental Division
Kysor Industrial Corporation
Marshall, Michigan

Salisbury has presented an excellent discussion of temperature and there seems little to disagree with in his paper. A few additional comments will be presented on how manufacturers of controlled environment equipment can be helpful to investigators.

The task of providing uniform guidelines for reporting conditions under which investigators perform their experiments is a formidable one. Such guidelines are necessary, but they should be kept rather flexible, because the needs of the various disciplines vary.

One purpose of the guidelines is to facilitate communication among investigators. It should be added that if manufacturers were more widely consulted, the usefulness of plant growth chambers would be greatly increased. The manufacturer is in business to assist customers and prospective customers. Hence investigators should not hesitate to contact several manufacturers before preparing specifications for new equipment. By so doing investigators will keep abreast of the art and learn more about equipment than can be given in brochures and price lists. All manufacturers prepare very detailed price lists, especially those that use the GSA Equipment Schedule, but often, some of the accessories in price lists are not compatible with one another and a researcher may

131

specify equipment that simply cannot be built. Communication via
telephone or letter with several manufacturers can often enable
the investigator to purchase a more flexible piece of equipment
than is possible after mere inspection of brochures and price
lists. Very often, modifications in basic design can be made that
are very inexpensive, but result in considerable flexibility.   At
other times a change that seems very simple to an investigator requires such extensive modification of the manufacturer's fixtures
or manufacturing methods that the modification is not cost-effective.

The investigator should also remember that, after he has received the instruments that he ordered, it is his obligation to
test the product, prior to initiation of research to be sure that
it lives up to the specifications. Once research is underway it
is the investigator's responsibility to regularly monitor the performance of his equipment, particularly between experiments, to
be sure that calibration is maintained and that the equipment is
operating up to its original specifications. Too many investigators assume that an instrument will maintain its calibration for
an indefinite period. Investigators should probably agree among
themselves on some simple calibration methods and equipment. It
is difficult to improve on a mercury thermometer - motor driven
psychrometer to obtain wet-bulb and dry-bulb measurements, that
provide both temperature and humidity measurements.

To obtain good equipment and to keep it running properly the
investigator should not overlook the services that are offered by
most manufacturers, either without charge or at very nominal cost.
Most manufacturers offer factory start-up and check-out of new
growth chambers for a very nominal fee. The investigator's request
for this service can be spelled out in specifications for new
equipment and a fee quoted by manufacturers can be part of the investigator's price evaluation. A nonfunctioning or poorly functioning growth chamber is bad advertising for manufacturers and

most of them will arrange for inspection and consultation about malfunctioning equipment, without cost to the user, even though the equipment may be beyond warranty. Modification kits are offered by most manufacturers to help upgrade equipment to a state-of-the-art condition.

As part of his purchase specification, the investigator might find it very useful to request of the manufacturer, at the time of delivery of new equipment, a set of "as built" specifications. The specifications would include, but not be limited to:

A. Quantity and manufacturer's designation as well as input wattage of each type of lamp utilized.

B. PAR reading at a distance of 12" from the lamps.

C. Type of air delivery (upward, downward, horizontal).

D. Barrier or no barrier- material used.

E. Symmetrical or asymmetrical air flow.

F. Remote or self-contained condensing units.

G. Modulated or cyclic flow of refrigerant through evaporator coil(s), etc.

In presenting a paper for publication, the investigator would, of course, present the guideline information and add that a complete physical description of the plant growth chamber used is available on request. Other investigators could contact the investigator to find out precisely what effect his choice of plant growth chambers might have had on the data obtained. It would then be possible to reproduce more closely the conditions under which the investigator conducted his work.

Finally, it is well to emphasize the paramount importance of communication between investigators and manufacturers. If the manufacturer is not aware of the investigator's problems with research equipment, he cannot help him even though he would like to do so.

TEMPERATURE: GUIDELINES

Lawrence R. Parsons

Department of Horticultural Science and
Landscape Architecture
University of Minnesota
St. Paul, Minnesota

Temperature, along with visible radiation, water and nutrition, is one of the major environmental factors affecting plant growth. Plant metabolism and rates of biochemical reactions are strongly dependent on temperature. The rates of many biological reactions will be increased by 2 to 3 times for each 10°C increase in temperature. Chilling or freezing temperatures severely limit the growing season of many crops. High temperatures at critical times during growth can cause flower abortion and reduce yield. Denaturation of enzymes can result from temperatures above or below certain levels. Because of the effect of temperature on saturated vapor pressure and the vapor pressure deficit, temperature also has a major effect on the rate of water loss.

Plants are seldom in thermal equilibrium with their environment. Salisbury pointed out that leaf to air temperature differences as large as 22°C sometimes occur at high altitudes. However, at lower radiation loads in the growth chamber, leaf temperatures are usually closer to air temperatures.

In developing guidelines for controlled environments, a major problem is to decide on the number of parameters to measure. For guidelines to receive widespread acceptance, measuring devices and procedures need to be readily available, relatively inexpensive and practical to use. It could be argued that one reason foot-candle measurements persisted as long as they have is that foot-candle meters meet these three requirements. Fortunately, many temperature measuring devices and procedures do meet these requirements.

135

RECOMMENDATIONS

The committee recommends that temperature in the chamber be measured with a shielded and aspirated (greater than 3 m s^{-1}) instrument. Salisbury's shielded thermocouple sensor is a good example of a simple, relatively inexpensive device for measuring temperature. At low radiation levels, shielding may not be necessary but at levels commonly used in controlled environments, shielding is recommended and at higher levels, it is necessary to avoid significant radiation errors. A sensor with a small heat capacity is desirable to obtain measurement of rapid temperature fluctuations. One with a large heat capacity will tend to average temperature changes and thus cannot monitor rapid temperature fluctuations. However, no guidelines are being proposed for the size of the sensor; if the sensor is adequately ventilated, sufficient precision should be obtained.

Units for measurement are °C which is consistent with SI terminology. In terms of where and when to measure, temperature measurements are similar to the other parameters discussed. The committee recommends that temperature be measured at the top of the plant canopy and averaged over the plant growing area. This is the region of the exposed leaves and highest photosynthetic rates. Measurement should be made with the chamber door closed. Because of air currents and differences in radiation at the edges and corners of the chamber, the average over the growing area is important. Temperature measurements within the canopy are not recommended because with different canopy densities and air flow rates, measurements in the canopy will vary in their relation to temperatures at the top of the canopy. If a light barrier is not present, vertical temperature gradients can become sizable. However, no recommendations are being made regarding monitoring vertical gradients in the chamber at this time. Although it would be useful to know leaf temperature, problems of precision of measurement, different species response, and leaf position would complicate the measurement.

Temperature measurement should be made hourly over the period of the study and continuous measurement is advisable. Hourly rather than daily measurements were considered necessary to make certain that short periods of low or high extremes be recognized and averaged. However, the more rapid temperature differences occurring with compressor cycling would not be separately reported. Hourly data would also be necessary to obtain a true picture of the average temperature to which plants were subjected. Data to be reported include the average of the hourly average values for the light and dark periods of the study along with the range of variation over the growing area. The range of variation refers to only the range in space across the chamber for the range in time would be included in the average reported. As was pointed out, variation across the chamber will probably exceed the precision and accuracy of the measuring instrument.

Soil temperature measurement is important but is often overlooked. Because soil temperature has such a marked effect on plant growth, water uptake, and various metabolic processes, it should be noted in growth chamber studies. Berry and Ulrich emphasized the importance of soil temperature on phosphorous uptake. Particularly with warm season crops, cold soil or application of cold irrigation water can cause wilting. A knowledge of soil temperature is critical because it may be similar to or quite different from the air temperature depending on the air flow and type of chamber.

The committee recommends measurement in °C in the center of a representative pot. For simplicity, measurement in one container is recommended because it is assumed that the variation in soil temperature over the different containers will be similar to the air temperature variation over the containers. The container can be moved to different locations in the chamber to determine the range of variability.

Hourly measurements should be monitored during the first 24 hours of the study and hourly measurements over the entire study are advisable. Hourly measurements for at least one day should be followed to make certain that the temperature recorded takes into account the slower temperature changes in soil with diurnal temperature fluctuations. Measurements made at any given time cannot be considered to represent average soil temperature. The average of the hourly average values for the (1) light and (2) dark periods for the first day (or over the period of the study if taken) would then be reported. The hourly average soil temperature for the first day is recommended because it is assumed that the difference between the air and soil temperature remains relatively constant over the study. This assumption may not hold true because the development of a dense canopy could alter the temperature difference between soil and air. However, because of the difficulty of obtaining continuous soil measurements, continuous measurements are not required.

One major advantage of growth chambers is that they can provide relatively stable and repeatable day and night temperatures. These recommendations have been kept simple in order to encourage the widespread use of a uniform procedure. Taken with care, temperature measurement should cause little difficulty even for the novice chamber user.

TEMPERATURE: DISCUSSION

WALTERS: The placement of temperature sensors in shielded-aspirated units should be questioned.   This only measures the air temperature at that unit.   The sensor can be placed nearly anywhere and if the investigator wants to regulate leaf temperature, the sensor should be placed on the leaf.

SALISBURY: Manufacturers should be concerned with regulating only air temperature, not leaf temperature.   There is too much variability in leaf temperature on a single plant for controlling a chamber by the leaf temperature.

WENT: I concur with Salisbury.   The large variability in temperature of different leaves on individual plants makes it impractical to use leaf temperature for control.

CAMPBELL: The importance of root temperature and its effect on rate of water uptake should not be neglected.

PALLAS: In our experiments wilting of snapdragons was caused by cool soil temperatures.

BERRY: The temperature of irrigation water should be recorded because of its effects on soil temperature and water uptake.

HELLMERS: It is desirable to precondition water or nutrients to the chamber temperature so there will be little change in temperature when irrigating.

KOLLER: The rate of change from day to night temperature should be reported to provide an accurate indication of the temperature for the plants.

TIBBITTS: This would add to problems in reporting.   Would not averaging hourly temperatures satisfy the need?   There is a difference in the rate of change of soil temperature in small versus large pots.   This should be considered in research.

PALLAS: To avoid temperature variation in small pots, we maintain the same day and night temperature in some studies.

McFARLANE: Because of the gradients in large and small pots, where should measurements of soil temperature be taken in the pot?

PARSONS: The recommended guideline is to measure temperature in the center of the pot. The pot could be moved to different locations to determine variability.

WENT: Because of sunlight in a greenhouse, soil temperature can be considerably different from air temperature and large pot-to-pot differences may occur.

HELLMERS: If investigators want to measure temperature where the roots are, they should remember that most of the roots are at the surface and bottom of the pot.

TIBBITTS: Uniform distribution of roots was found in a peat vermiculite mix when compaction was avoided with automatic systems with drip watering.

KRIZEK: The type of pot should be reported because there is a difference in the amount of evaporation and hence cooling of porous versus nonporous pots. Thus there is also a soil temperature interaction with humidity when porous pots are used.

READ: The measurement of leaf temperature should be taken to provide an integrated index of all environmental effects.

PARSONS: This would be desirable but difficulties in measuring leaf temperature led to a recommendation for reporting only air temperature at this time.

HUMIDITY

Glenn J. Hoffman

U.S. Salinity Laboratory
Riverside, California

INTRODUCTION

Predominant in the evolution of terrestrial plants is their adaptation to grow outside an aqueous environment. The capacity to maintain suitable hydration within physiologically active tissues during periods of desiccation is paramount to plant survival. Vascular plants have developed stomata, which control the rate of water loss to the atmosphere and regulate exchange of carbon dioxide and oxygen between the plant and atmosphere, thereby influencing plant growth. The epidermis and cuticle together constitute an effective barrier to water loss through the remainder of the leaf surface. Because these adaptive mechanisms for coping with unsaturated aerial environments are well known, investigators have often disregarded the influence that atmospheric humidity has on plant growth.

Until recent years, little information was available on the influence of humidity on plant growth, even less on the yield of crop plants. Such limited information was no doubt the result of a lack of humidity control in early environmental facilities and the low importance placed on humidity by scientists pioneering in controlled-environment studies (i.e., Went, 1957). Early data did indicate, however, that very dry air causes high

rates of transpiration, inducing plant water deficits that restrict growth. This was particularly true for plants with limited root systems or for plants grown under conditions of low water
supply. Air that was almost saturated also was shown to have
deleterious effects. Apart from favoring pests and diseases,
high humidity sometimes reduced plant growth (Freeland, 1936;
Winneberger, 1958; Nieman and Poulsen, 1967). Occasionally,
plants grown under high humidities were morphologically abnormal,
suggesting hormone imbalance (Pareek, Sivanayagam, and Heydecker,
1969).

Recent studies in controlled environments have shown that
growth and yield of crops can differ as humidity varies (Hoffman,
1973; Ford and Thorne, 1974; O'Leary, 1975). Such investigations
emphasize the importance of careful regulation of humidity in
controlled environment research on plants.

The principal intent of this paper is to propose guidelines
for control of atmospheric humidity. First, however, the sensitivity of plants to humidity will be discussed before guidelines
for the degree and accuracy of measurement and control will be
outlined. This discussion includes an estimate of: (1) the range
within which humidity may fluctuate without influencing plant
response, and (2) the range over which humidity produces measura-
ble effects on plants. Potential adverse effects of extremely
high humidity will also be described. Brief descriptions of the
various types of humidifiers, dehumidifiers, and humidity sensors
that may be used in environment-controlled facilities will also
be given.

TERMINOLOGY

The atmosphere, a mixture of gases, is almost constant in
composition except for variations in the quantity of water vapor.
Water vapor constitutes less than 2% of the atmosphere, with the
lowest concentrations occurring in cold or arid regions and the

highest in the moist tropics. According to Dalton's law of
partial pressures, the pressure exerted by water vapor is independent of the pressure exerted by other atmospheric gases. Furthermore, provided conditions are such that no condensation or
evaporation is occurring in the volume of air under consideration,
the behavior of water vapor is similar to that of other gases,
which can be represented approximately by the equation of state
for an ideal gas. The vapor pressure of water (e in mb) under
these conditions can be represented by

$$e = \rho_v \, RT/M_w,$$

where ρ_v is the density of water vapor (kg m^{-3}), R is the universal gas constant (0.0831 mb m^3 mol^{-1}K^{-1}), T is the absolute
temperature of the water vapor and is assumed to be the temperature of the air, and M_w is the molecular weight of water (0.018
kg mol^{-1}). The density of water vapor is referred to as the
absolute humidity of the atmosphere.

The kinetic theory of gases indicates that evaporation occurs
when molecules in a liquid attain sufficient energy to overcome
mutual attractive forces and escape from a free surface. When
the number of water molecules leaving the surface is equal to
those returning per unit time, the atmosphere is said to be
saturated with water vapor. The saturation vapor pressure, e_s,
is a unique function of temperature. The rapid increase of the
saturated water vapor pressure of air with rise in temperature
is illustrated in Fig. 1 and can be expressed numerically by:

$$\log e_s = 0.026404 \, T + 0.82488$$

where T is the air temperature in °C (Rosenberg, 1974).

For an air space not saturated with water vapor ($e < e_s$), the
degree of unsaturation can be expressed in terms of the saturation
deficit or relative humidity. The saturation deficit, Δe, is
given by

$$\Delta e = e_s - e,$$

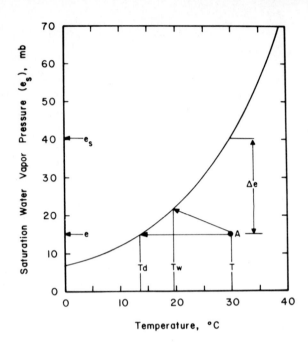

FIGURE 1. The saturation vapor pressure of water (e$_s$) as a function of air temperature (T). Also illustrated are dew point (T$_d$) and wet-bulb (T$_w$) temperatures and saturation deficit (Δe).

provided that both e and e$_s$ are measured at the same temperature. Relative humidity, H$_r$, is the ratio e/e$_s$ expressed as a percentage

$$H_r = 100 \ e/e_s.$$

Neither the ratio nor the difference, however, allows e or e$_s$ to be calculated. Thus, besides H$_r$ or Δe, some other parameter, such as air temperature, must be given before the humidity condition of air is fully defined.

The dew point temperature, T$_d$, also a measure useful in describing the humidity level, is the temperature at which the vapor in a parcel of air would reach saturation. This can be illustrated by reference to Fig. 1 where a parcel of air whose humidity is represented by A is cooled at constant pressure without any gain or loss of water vapor. The horizontal line to the left of A represents the path of this change. It intersects the

saturation vapor pressure curve at the highest temperature at
which air will be saturated with water vapor. This temperature
is called the dew point, T_d, because dew would condense on any
surface in contact with the vapor if the surface were cooled
below T_d.

The wet-bulb temperature, T_w, another useful measure for
describing the humidity level, can also be illustrated in Fig. 1
by considering the cooling of air as it passes over a wet surface.
Again, if point A represents the initial state of an air parcel,
the evaporation of water into an air stream will both increase
the vapor pressure and reduce the temperature of the air parcel
by extracting from it the latent heat required for evaporation.
Such a change can be represented in the illustration by the path
that inclines upwards and to the left of A in Fig. 1. This path
intersects the saturation water pressure curve at a particular
temperature, provided the air flow rate is greater than about
5 m s^{-1}. A small wet surface achieves a temperature which
decreases insignificantly with further increases in flow rate,
provided this small wet surface is, to some extent, thermally
isolated from its surroundings. This temperature is defined as
the wet-bulb temperature, T_w, for an air parcel at state A.

The capacity of air to hold water vapor increases rapidly as
temperature increases, approximately doubling for every 10°C
rise in the temperature range of plant growth (Fig. 2). Thus,
air contains significantly different amounts of water vapor at
different air temperatures although the relative humidity is the
same. The amounts of water held in air can be closely approximated by e over the temperature range of interest because, from
Eq. (1), the relationship of e and ρ_v is altered only slightly
by changes in the absolute temperature. An atmosphere 60%
saturated at 20°C (e = 13.3 mb) contains only half as much water
vapor as an atmosphere 60% saturated at 30°C (e = 24.2 mb).
More significantly, the capacity of the air to hold additional
water vapor is much lower at 20°C (Δe = 22.2 - 13.3 = 8.9 mb)

than at 30°C (Δe = 41.4 - 24.2 = 17.2 mb). Therefore, water will
evaporate more rapidly at 30°C than at 20°C even though the
relative humidity is identical. The saturation deficit of water
vapor pressure (Δe) at a given temperature indicates the difference between the amount of water vapor present in the atmosphere
and the maximum that could be held at that temperature; equal
Δe indicates a similar rate of evaporation, other factors being
equal. An additional advantage of saturation deficit is that it
is a more sensitive indicator of the water vapor conditions and
varies over a wider range with temperature change than does
relative humidity. As an example, the vapor pressure at 20°C
and a relative humidity of 60% is 13.3 mb (Fig. 2), whereas the
same vapor pressure at 30°C results in a reduction of relative
humidity to 32% (a 2-fold change). Saturation deficit, on the
other hand, changes from 8.9 to 28.1 mb (a 3-fold change).

Because saturation deficit and transpiration are closely
related, saturation deficit is preferable to relative humidity
when considering plant responses to atmospheric humidity. For
clarity, relative humidity may also be given. With either,
temperature must also be reported. When referring to control
and operation of environmental chambers, dew point temperature
or wet-bulb temperature should be emphasized because of their
greater accuracy in sensing and controlling atmospheric humidity.

PLANT RESPONSE TO INCREASED HUMIDITY

The influence of atmospheric humidity on growth of 26 crops
is summarized in Table 1. The main intent of the table is to
show whether a reduction in saturation deficit (or an increase
in relative humidity) influenced yield, shoot weight, root weight,
leaf area, or plant height. A separate column indicates whether
an atmosphere nearly saturated with water vapor (small saturation
deficit or high atmospheric humidity) has been shown to be

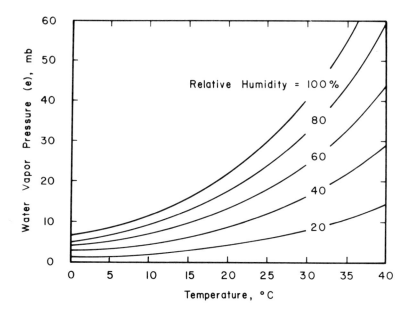

FIGURE 2.   Atmospheric vapor pressure of water as a function of relative humidity and air temperature.

detrimental to plant growth, flowering, or seed production. The general conclusions that can be drawn from the literature are presented, together with comments on specific crops.

Shoot weight of all the ornamental plants cited was stimulated by increases in atmospheric humidity. Reducing Δe from 18 to 10 mb significantly increased shoot weight, leaf area, and plant height of ageratum, marigold, and petunia plants. A further reduction of Δe to 3 mb was of limited benefit. For chrysanthemum, decreasing Δe from 16 to 5 mb increased shoot weight.

All the vegetable crops tested benefited from an increase in humidity except pea, which did not respond. All four studies with bean indicated that reducing Δe from more than 15 to less than 5 mb increased all the growth parameters measured. Reducing Δe from more than 10 to less than 5 mb was also beneficial for kale, lettuce, and tomato. Decreasing Δe from 28 to 6 mb significantly increased pepper yields.

TABLE 1. *Plant Response to a Significant Reduction in the Saturation Deficit (Significant Increase in Relative Humidity). B indicates a Benefit (Increased Growth) from a Reduced Saturation Deficit, N denotes No Effect or that the Effect is Variable, D indicates that a Reduced Saturation Deficit is Detrimental (Reduced Growth). An Asterisk denotes that the Effect is Statistically Significant at α < 0.05.*

Plant	Reduction in saturation deficit					Effect of a small saturation deficit (Δe < 3 mb)	Reference
	Yield	Shoot weight	Root weight	Leaf area	Plant height		
Ageratum (Ageratum houstonianum)	–	B*	–	B*	B*	B	Krizek, Bailey, Klueter (1971)
Apple (Malus sylvestris)	B	B*	–	–	–	–	Tromp and Oele (1972)
Barley (Hordeum vulgare)	D	N	B	–	D	D	Hoffman and Jobes (1978)
Bean (Phaseolus vulgaris)	–	–	–	–	–	D	Freeland (1936)
	B	B	–	–	–	–	Hoffman and Rawlins (1970)
	–	B	–	–	–	–	Lunt, Oertli, and Kohl (1960)
	–	–	–	–	–	–	Nieman and Poulsen (1967)
	B	B	–	B	–	D	O'Leary and Knecht (1971)

Plant	Reduction in saturation deficit					Effect of a small saturation deficit ($\Delta e < 3$ mb)	Reference
	Yield	Shoot weight	Root weight	Leaf area	Plant height		
Bean (continued) (Phaseolus vulgaris)	–	B*	B*	B*	–	–	Prisco and O'Leary (1973)
Beet, garden (Beta vulgaris)	B*	B	N	–	–	–	Hoffman and Rawlins (1971)
Brussels-sprouts (Brassica oleracea gemmifera)	–	–	–	–	–	D	van Marrewijk and Visser (1978)
Cabbage (Brassica oleracea capitata)	–	–	–	–	–	D	van Marrewijk and Visser (1978)
Cacao (Theobroma cacao)	–	D*	D*	D*	B*	D	Sale (1970)
Chrysanthemum (Chrysanthemum)	–	B	–	–	–	–	Lunt, Oertli, and Kahl (1960)
Coleus (Coleus blumei)	–	–	–	–	–	D	Freeland (1936)

TABLE 1. (continued).

Plant	Reduction in saturation deficit					Effect of a small saturation deficit ($\Delta e < 3$ mb)	Reference
	Yield	Shoot weight	Root weight	Leaf area	Plant height		
Corn (Zea mays)	B	B	-	-	-	-	Hoffman (1973)
	N	D*	D	-	B	-	Hoffman and Jobes (1978)
	D	N	N	B	B	-	McPherson and Boyer (1977)
	-	B	-	-	B	-	Pareek, Sivanayagam, and Heydecker (1969)
Cotton (Gossypium hirsutum)	B*	B	D	B	B	D	Hoffman and Rawlins (1971)
	-	-	-	-	-	D	Nieman and Poulsen (1967)
Kale (Brassica oleracea acephala)	-	B*	B	B*	-	B	Ford and Thorne (1974)
Lettuce (Lactuca sativa)	-	B*	-	B*	-	-	Tibbitts and Bottenberg (1974)
Marigold (Tagetes erecta)	-	B*	-	-	B*	-	Krizek, Bailey Klueter (1971)

Plant	Reduction in saturation deficit					Effect of a small saturation deficit ($\Delta e < 3$ mb)	Reference
	Yield	Shoot weight	Root weight	Leaf area	Plant height		
Onion (Allium cepa)	N	–	D	–	–	D	Hoffman and Rawlins (1971)
Pea (Pisum sativum)	N	–	–	–	N	–	Nonnecke, Adedipe, and Ormrod (1971)
Peanut (Arachis hypogaea)	B	B	–	–	–	–	Lee, Ketring, Powell (1972)
	B	B	–	–	–	–	Fortanier (1957)
Pepper (Capsicum annuum)	B	–	–	–	N	D	Baer and Smeets (1978)
	B*	B	–	–	–	–	Hoffman (1975)
Petunia (Petunia hybrida)	–	B*	–	B*	–	B	Krizek, Bailey Klueter (1971)
Radish (Raphanus sativus)	B	B	–	–	–	–	Hoffman and Rawlins (1971)
Safflower (Carthamus tinctorius)	D	–	–	–	–	–	Zimmerman (1972)

TABLE 1. (continued).

Plant	Reduction in saturation deficit					Effect of a small saturation deficit ($\Delta e < 3$ mb)	Reference
	Yield	Shoot weight	Root weight	Leaf area	Plant height		
Soybean (Glycine max)	–	–	–	N	–	–	Beardsell and Mitchell (1973)
	B*	B	–	–	–	–	Woodward and Begg (1976)
Sugarbeet (Beta vulgaris)	–	B*	B	B*	–	B	Ford and Thorne (1974)
Sunflower (Helianthus annuus)	–	–	–	–	–	D	Freeland (1936)
	–	B	B	–	–	–	Demidenko and Golle (1939)
Tomato (Lycopersicon esculentum)	–	–	–	–	–	D	Freeland (1936)
	–	B	–	–	–	–	Nightingale and Mitchell (1934)
	–	B*	B*	–	–	B	Swalls and O'Leary (1975)
	–	–	–	–	B	–	Went (1957)

Plant	Reduction in saturation deficit					Effect of a small saturation deficit (Δe < 3 mb)	Reference
	Yield	Shoot weight	Root weight	Leaf area	Plant height		
Wheat (Triticum aestivum)	N	N	–	–	–	–	Campbell, McBean, and Green (1969)
	–	B,N	B,N	B,N	–	–	Ford and Thorne (1974)
	B	B*	B	–	N	–	Hoffman and Jobes (1978)

153

The yield of root crops either was improved or not influenced by humidity. A Δe of less than 5 mb increased yield and shoot weight when compared with a Δe of 13 mb for both garden beet and radish. Onion yields at the same two values of Δe did not differ, although root weight was decreased at the lower Δe value.

Shoot weight of all sugar, fiber, and oilseed crops evaluated, except safflower, increased under high humidity. Reducing Δe from 31 to 16 mb increased cotton yield and all aspects of growth measured except root growth. Either extremely high ($\Delta e < 10$ mb) or low ($\Delta e > 35$ mb) humidity, however, caused cotton flowers to be infertile, and yields were negligible. For peanut, flowering and shoot weight increased at a Δe of 2 mb when compared with 21 mb. Low humidity (Δe of either 11 or 22 mb as compared to 8 mb) reduced soybean yield as a result of reduced pod numbers caused by floret abortion. Shoot weight of soybean was also less at a larger Δe. Reducing Δe from 8 mb or more to less than 4 mb increased leaf area and shoot and root weights of sugarbeet. Likewise, a lower Δe increased the weight of sunflower shoots and roots. Increasing humidity for one day during flowering reduced yield of safflower over a range of air temperatures.

Increased humidity was of dubious benefit to cereal crops. Decreasing Δe from 14 to 3 mb reduced grain yield and plant height of barley. Shoot weight of barley was not influenced, but root weight increased. Plant height and leaf area of corn were increased by high humidity but the effect on yield and shoot weight was not consistent among experimenters, even though Δe varied from more than 20 to less than 5 mb. Experiments with wheat also were inconsistent, with some tests showing increased weight after reducing Δe from 10 to 5 mb, and others showing no effect.

Only two tree crops were studied and they differed markedly in response to humidity. Shoot weight, root weight, and leaf area of young cacao trees were reduced as Δe decreased from 16 to

9 mb and from 9 to 2 mb. Only plant height was greater at high
humidity. As was typical of most other crops, shoot weight and,
to a lesser degree, fruit weight of apple improved as Δe
decreased from 11 to 6 mb.

In summary, of the 26 crops tested, only six (barley, cacao,
corn, onion, pea, and safflower) did not show increased growth
from increased atmospheric humidity. Growth was often increased
when Δe was changed from about 10-15 mb to about 5 mb. Statistically significant increases in growth, in spite of the limited
number of replications in these studies, were typical if the
saturation deficit was reduced by about 10 mb, provided that
extremely high humidities were avoided.

EFFECTS OF EXTREMELY HIGH HUMIDITY

In addition to the growth-limiting and undesirable effects of
low atmospheric humidity, some potentially adverse physiological
effects are associated with extremely high humidity. One of
these effects is heat damage (Kinbacher, 1962). As humidity
increases and transpirational cooling decreases, heat loss from
leaves by conduction and convection becomes important, but
depends on the rate of air movement across the leaf. Thus, if
high humidity and low air movement occur simultaneously, heat
damage is likely to occur. Positive correlations between leaf
temperature and atmospheric humidity have been shown for numerous
crops (Carlson, Yarger, and Shaw, 1972; Barrs, 1973; Forde, Mitchell, and Edge, 1977).

Another potentially harmful effect of high humidity is
increased injury to plants by air pollutants. Plants are
injured more by atmospheric pollutants at high than at low
humidity because stomatal conductance is higher. Foliar injury
to tobacco and bean from ozone is highly correlated with stomatal
conductance (Otto and Daines, 1969; Dunning and Heck, 1977).
This is also true for sulfur dioxide injury (O'Gara, 1956).

High humidity may reduce the translocation of ions to shoots
because of reduced transpiration. Although concentration of
some ions in leaves was lowered under high humidity conditions
(Freeland, 1936; Demidenko and Golle, 1939; O'Leary and Knecht,
1972; Tromp and Oele, 1972), reduction of ions to injurious levels
has not been reported. There are indications, however, that
when humidity is high, reduced calcium transport is associated
with physiological disorders of tomato and apple fruits (Wiersum,
1966; Tromp and Oele, 1972).

Hormones such as gibberellins and cytokinins apparently are
synthesized in the roots and transported by the transpiration
stream to the leaves. Although hormonal imbalance caused by
reduced transport under high humidity has not been demonstrated,
the formation of adventitious roots on stems (Hughes, 1966)
and abnormal flower development as in cotton (Hoffman and Rawlins,
1971) at high humidity suggest hormonal changes.

Studies in which extremely high humidity was detrimental to
plant growth are cited in Table 1. For this tabulation,
extremely high humidity indicates a saturation deficit of less
than approximately 3 mb. Most crops studied were affected
adversely at extremely high humidities. When very high humidity
was shown to be beneficial, the primary measure was shoot weight,
as in ageratum, kale, petunia, and sugarbeet. However, in other
crops, including bean, cacao, coleus, cotton, and sunflower,
shoot weight was reduced by high humidity. Instances where
extremely high humidity was detrimental often were associated
with low flower fertility as in Brussels sprouts, cabbage, cacao,
cotton, and pepper.

HUMIDITY CONTROL EQUIPMENT

Control of atmospheric humidity requires equipment to sense
changes and to humidify and dehumidify. The equipment must have
the necessary capacity, response time, and accuracy to maintain

humidity at the desired levels without unacceptable alterations
of other environmental parameters. The requirements of certain
studies may narrow the choice of instruments to those capable of
responding to proportional controllers.

A mandatory requirement, often neglected in environmental
studies, is a set of independent and preferably more precise
sensors to verify the accuracy of the sensors controlling the
air-conditioning equipment. The same sensor must not be used to
control and record humidity without verification that the sensor
is accurate and is maintaining the desired level within the plant
canopy. If a sensor that responds directly to changes in relative humidity operates the humidity-control equipment, it is
desirable to use an instrument that employs another principle,
such as detection of dew point temperature or water vapor concentration, to verify the accuracy and calibration of the
humidity sensor. Depending on the humidity control desired and
the controlling sensor, the frequency of accuracy checks may
vary from continuously to weekly.

Humidification

When plants are young, when plant population is sparse, or
when temperatures are low, large quantities of water are frequently required to maintain higher than ambient humidity levels.
Several techniques are available for adding water vapor to air;
some are more easily controlled than others and lend themselves
to proportional control where the amount of correction is proportional to how much the environment is out of control. In the
past, saturated salt or acid solutions were placed in small
plant chambers to maintain the humidity at a level prescribed by
the vapor pressure of the chosen solution. The requirement for
large quantities of water vapor in most controlled-environment
facilities makes this highly impractical except in very small
chambers. Sprayers or atomizers that direct a fine mist of water

into the air stream are used commercially in some environmental
chambers. The fine droplets evaporate in the air stream,
increasing the water vapor of the air. Evaporation is most
effective when the spray is directed perpendicular to or directly
against the air stream. The capacity and response of this type
of equipment depends on minimizing the number of droplets that
must be recycled because they are too large to evaporate. Some
commercial chambers provide water baths for humidification. Air
is blown over the bath and then directed into the chamber. The
amount of humidification provided is a function of the time the
air is exposed to the water surface, the water temperature, and
the air turbulence over the water bath. Increasing the temperature of the water bath has limitations for increasing humidity
because heat is also added to the air, necessitating additional
cooling to maintain temperature which, in turn, tends to reduce
the vapor pressure of the air. Steam injection is the most
efficient humidification system because it affects temperature
least. Response is fast, and the steam injection rate can be
controlled in a proportioning manner.

Dehumidification

Close control of humidity requires a dehumidification system
to offset evapotranspiration, the high humidity of incoming
ambient air, and the humidifier. Normal operation of an environment-controlled facility equipped with cooling coils provides
some dehumidification as water is condensed on the cold coil.
For minimal changes in humidity from the ambient condition or
when there is a small evapotranspiration load in the chamber,
this may be adequate. Of course, the amount of cooling required,
and hence the amount of dehumidification, increases with increasing heat loads from lights and other sources in the chamber.
Humidity is often controlled by placing two cooling coils in
parallel. One acts as a dehumidifier, the other cools for

temperature control. The dual cooling system can provide
modulated control by passing a fraction of the conditioned air
through the humidity control system while the remainder passes
through the temperature control coils. The fraction of air
passing over the dehumidification coil is adjusted to achieve
the desired final humidity. Heat is added after the dual cooling
system to achieve the desired air temperature in the chamber. To
prevent the water condensed onto cooling coils from freezing,
the surface temperature of the coil must be maintained above
2°C. Thus, at an air temperature of 15°C, for example, the
lowest relative humidity possible is about 45%. For lower
humidity, the refrigerant coils must be periodically defrosted
while maintaining adequate control with a second cooling system.
Chemical driers may be used to reduce humidity below that
obtainable with cooling coils. The major problem with chemical
driers is the large capacity required to remove the water still
present in the air after it passes over precooling coils.
Frequently, two chemical driers are required so that one can
be regenerated while the other is dehumidifying.

Sensors

Several types of commercial instruments are available for
monitoring atmospheric humidity. The most common and perhaps
cheapest instrument is the psychrometer. It consists of two
identical temperature-sensitive elements (i.e., thermometers,
thermocouples, resistive elements), one of which is covered with
a moistened wick. Evaporation cools the wet element in proportion to the saturation deficit of the air. The temperature of
this wet element and the air temperature taken from the dry
element can be converted to relative humidity or saturation
deficit (Table 2). Air must move over the psychrometer at a
speed of at least 5 m s^{-1} for consistent readings. Precise
humidity measurements normally cannot be obtained without

TABLE 2. Conversion of Wet- and Dry bulb Temperatures to Saturation Deficit or Relative Humidity.

Wet bulb Depression $(T-T_w)$ °C	Saturation deficit (Δe), mb — Dry bulb temperature (T), °C						Relative humidity (H_r), % — Dry bulb temperature (T), °C					
	15	20	25	30	35	40	15	20	25	30	35	40
0.0	0.0	0.0	0.0	0.0	0.0	0.0	100	100	100	100	100	100
0.5	0.8	0.9	1.2	1.6	1.6	2.2	95	96	96	96	97	97
1.0	1.6	2.0	2.4	2.8	3.3	4.4	90	91	92	93	94	94
1.5	2.5	2.9	3.6	4.4	5.4	6.6	85	87	88	89	90	91
2.0	3.3	3.8	4.8	5.7	7.1	8.8	80	83	84	86	87	88
2.5	4.1	4.9	5.7	6.9	8.7	11.0	75	78	81	83	84	85
3.0	4.8	5.8	6.9	8.5	10.4	13.2	71	74	77	79	81	82
3.5	5.6	6.6	7.8	9.7	12.0	14.7	66	70	74	76	78	80
4.0	6.4	7.5	9.0	10.9	13.6	16.9	61	66	70	73	75	77
4.5	7.1	8.4	9.9	12.1	15.3	19.1	57	62	67	70	72	74
5.0	7.9	9.1	11.1	13.3	16.9	20.6	52	59	63	67	69	72
5.5	8.5	10.0	12.0	14.5	18.0	22.8	48	55	60	64	67	69
6.0	9.2	10.9	12.9	15.7	19.6	24.3	44	51	57	61	64	67
6.5	9.9	11.5	13.8	17.0	21.2	26.5	40	48	54	58	61	64
7.0	10.5	12.4	15.0	18.2	22.3	27.9	36	44	50	55	59	62
7.5	11.2	13.1	15.9	19.4	24.0	30.1	32	41	47	52	56	59
8.0	12.0	14.0	16.8	20.2	25.1	31.6	27	37	44	50	54	57
8.5	12.6	14.6	17.6	21.4	26.7	33.8	23	34	41	47	51	54
9.0	13.1	15.5	18.5	22.6	27.8	34.6	20	30	38	44	49	53
9.5	13.8	16.2	19.1	23.4	28.9	36.0	16	27	36	42	47	51
10.0	14.4	16.8	20.0	24.6	30.5	38.2	12	24	33	39	44	48
11.0	15.4	18.2	21.8	26.2	32.7	41.1	6	18	27	35	40	44
12.0		19.5	23.3	28.3	34.9	44.1		12	22	30	36	40

Wet bulb depression $(T-T_w)$ °C	Saturation deficit (Δe), mb						Relative humidity (H_r), %					
	Dry bulb temperature (T), °C						Dry bulb temperature (T), °C					
	15	20	25	30	35	40	15	20	25	30	35	40
13.0		20.8	24.8	30.3	37.6	47.1		6	17	25	31	36
14.0			26.3	31.9	39.8	49.3			12	21	27	33
15.0			27.8	33.5	41.4	52.2			7	17	24	29
16.0				35.1	43.6	54.4				13	20	26
17.0				36.7	45.8	56.6				9	16	23
18.0					47.4	58.8					13	20
19.0					49.0	61.0					10	17

Atmospheric pressure assumed to be 1013 mb.

161

several readings of the wet- and dry bulb elements because of
changes induced by heating and cooling cycles. Psychrometers
are also subject to error if the water is not pure, or if the
wick is not clean or does not completely cover the sensing
element. If all potential errors are avoided, the accuracy of
the humidity measurement depends on the accuracy of the temperature-sensing elements. If the temperature-sensing elements are
accurate to within 0.2°C, then Δe is known within 0.5 mb.

Hygroscopic elements are reasonably inexpensive and provide a
simple method of sensing humidity. Common materials used in such
elements are human hair, animal membrane, wood products, or
nylon, all of which expand and contract with changes in water
absorption. Unfortunately, no material has been found that will
consistently reproduce its action over an extended period. Such
devices require initial calibration and frequent recalibration,
especially if they are exposed to humidity extremes. The hair
humidistat has low sensitivity and considerable lag, and the
precision in measuring Δe is at most 1 mb.

Several types of instruments permit an absolute rather than a
relative measurement by determining the dew point temperature.
One such instrument consists of a tubular wick impregnated with
lithium chloride and mounted over a metal tube that is electrically insulated from the wick. Two parallel wires are wound on
the wick and connected to an alternating current source. The
lithium chloride absorbs water from the air and becomes conductive, allowing current to pass between the two wires through the
lithium chloride layer. The current generates heat, which tends
to evaporate the water and leads to a reduction in conductivity
and current. At equilibrium the temperature of the lithium
chloride produces a partial pressure of water that just equals
the vapor pressure over a standard lithium chloride solution.
Temperature is measured by a thermistor or similar temperature
sensor. The vapor pressure-temperature relationship of lithium

chloride is well known, so the output can be calibrated directly
in dew-point temperature. Water deposits on the element from
condensation or sprays change the calibration drastically and
thus frequent recalibration may be required. Dew point temperature can also be measured by detecting with an optical system
when the reflectance of a mirrored surface changes because of
condensation. Typically, an air sample is drawn over the mirrored
surface and cooled until vapor condenses. The instrument stabilizes at a new humidity level within a minute of any change.
Readings of air temperature and the dew point temperature can be
converted easily to saturation deficit or relative humidity
(Table 3). Infrared gas analyzers are the most precise instruments currently used to monitor humidity. The detectors are
large, but are ideally suited for remote operation by drawing an
air sample through the instrument. To achieve the precision
possible, Δe within 0.1 mb, calibration must be checked continually.

Adequate atmospheric humidity control for environmental
studies on plant growth normally requires equipment to add and
extract water vapor. For close control, this equipment is
typically operated by dew-plint or wet-bulb temperature sensors
that initiate proportional control.

GUIDELINES FOR HUMIDITY CONTROL

Guidelines are always subject to special circumstances that
negate them. Nevertheless, the following guides are proposed
for control and measurement of atmospheric humidity in environment-controlled chambers for plant growth studies. The guidelines should be applicable at temperatures of 15 to 40°C, the
range in which most plant growth studies are conducted. The
recommended humidity equipment can achieve the desired humidity
control if properly located, sized, and maintained. In addition

TABLE 3. Conversion of Dew Point Temperatures to Saturation Deficit or Relative Humidity.

Dew point depression $(T-T_d)$ °C	Saturation deficit (Δe), mb — Dew point temperature (T_d), °C					Relative humidity (H_r), % — Dew point temperature (T_d), °C				
	0	10	20	30	40	0	10	20	30	40
0.0	0.0	0.0	0.0	0.0	0.0	100	100	100	100	100
0.5	0.3	0.5	0.7	1.2	2.3	96	96	97	97	97
1.0	0.5	0.8	1.4	2.6	3.9	93	94	94	94	95
1.5	0.7	1.2	2.2	4.0	6.4	90	91	91	91	92
2.0	1.0	1.6	3.0	5.1	8.3	87	88	88	89	90
2.5	1.2	2.1	3.6	6.1	10.2	84	85	86	87	88
3.0	1.5	2.6	4.5	7.7	13.2	81	82	83	84	85
3.5	1.8	3.0	5.5	9.0	15.4	78	80	80	82	83
4.0	2.1	3.6	6.2	10.3	17.8	75	77	78	80	81
4.5	2.4	4.2	6.9	12.2	20.2	72	74	76	77	79
5.0	2.7	4.6	7.8	13.6	22.8	70	72	74	75	77
6.0	3.2	5.6	9.5	16.8	28.4	66	68	70	71	73
7.0	4.0	6.8	11.5	19.7	33.6	61	63	66	68	70
8.0	4.6	7.9	13.6	23.5	40.4	57	60	62	64	66
9.0	5.4	9.2	16.0	27.0	46.7	53	56	58	61	63
10.0	6.1	10.4	18.2	31.6	52.2	50	53	55	57	61
11.0	6.8	12.0	20.6	35.1		47	49	52	55	
12.0	7.7	13.2	23.2	39.8		44	47	49	52	
13.0	8.6	14.9	26.1	44.9		41	44	46	49	
14.0	9.6	16.6	28.7	49.5		38	41	44	47	
15.0	10.5	18.2	31.6	55.6		36	39	42	44	
16.0	11.5	20.0	35.3	61.1		34	37	39	42	

Dew point depression $(T-T_d)$ °C	Saturation deficit (Δe), mb					Relative humidity (H_r), %				
	Dew point temperature (T_d), °C					Dew point temperature (T_d), °C				
	0	10	20	30	40	0	10	20	30	40
17.0	12.6	21.9	38.7	67.1		32	35	37	40	
18.0	13.8	24.0	42.4	73.7		30	33	35	38	
19.0	15.0	26.2	47.4	80.7		28	31	33	36	
20.0	16.4	28.7	50.0	88.4		26	29	32	34	
22.0	19.2	33.7	59.7			23	26	28		
24.0	22.2	39.5	69.2			21	23	26		
26.0	26.0	45.7	81.1			18	21	23		
28.0	30.1	52.8	93.8			16	19	21		
30.0	34.7	61.0	108.5			14	17	19		

Atmospheric pressure assumed to be 1013 mb.

to the dew point or wet bulb temperature readings used for
control, periodic calibration checks must be made with a precise
humidity measuring instrument such as a dew point hygrometer or
infrared gas analyzer. During critical experiments, several
checks may be required each day. Although the placement of the
humidity sensor within a chamber is not as critical as for other
environmental monitors, the sensor should be located near the
plant canopy, yet shielded from radiation. Of course, any
ventilation requirement must be met.

Saturation deficits (Δe) of less than 3 mb should be avoided,
unless such values represent the specific humidity range of
interest. At 25°C, a relative humidity above 90%, particularly
during the day, is not recommended (Fig. 3). As discussed
previously, at extremely high humidities plant growth may be
reduced, flowers are frequently infertile, nutrient and hormone
imbalance may occur, and diseases or pests are more likely.

If the intent of an experiment is to study effects of
environmental factors other than humidity, questions often are
asked about the humidity level that is optimal for crop production. The data reviewed here indicates that most plants grow
well when Δe is maintained between 5 and 10 mb. At 25°C, the
optimum relative humidity range would be 65 to 85% (Fig. 3).
Downs (1975) recommended that a saturation deficit equal to a
relative humidity of 70% at an air temperature of 25°C (Δe = 9 mb)
be maintained when humidity control is not a critical aspect of
the experiment. Most environment-controlled facilities with
provision for humidity control can maintain humidity in this
range with an accuracy at 25°C of ±1.0 to 1.6 mb of vapor
pressure (±3 to 5% relative humidity). This should be adequate
for all except the most meticulous experiments. Facilities
without humidity control frequently have humidity levels below
this range and less than maximum plant growth should be expected
for most crops.

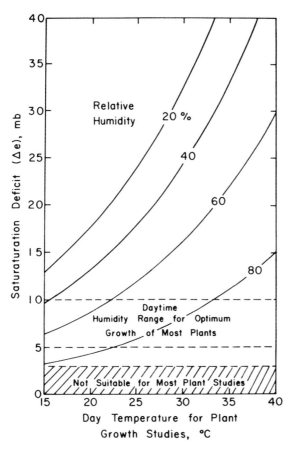

FIGURE 3. Saturation deficit as a function of temperature and relative humidity. Humidity guidelines are also illustrated.

Experimentation on the influence of humidity on plant growth typically shows that differences in shoot growth occur when the saturation deficit differs by more than 5 mb among treatments. Humidity treatments differing by about 10 mb generally induce significant growth differences, even with relatively few replications. Thus, cultural and management practices that decrease Δe by more than 5 mb are potential tools for increasing production of many crops and warrant experimental verification.

REFERENCES

Baer, J., and Smeets, L. (1978). Effect of relative humidity on
    fruit set and seed set in pepper (*Capsicum annuum L.*). *Neth.
    J. Agr. Sci. 26*, 59-63.

Barrs, H. D. (1973). Controlled environment studies of the
    effects of variable atmospheric water stress on photosynthesis,
    transpiration, and water status of *Zea mays* L. and other
    species. *In* "Plant Response to Climatic Factors" (R. O.
    Slatyer, ed.), pp. 249-258. *Proc. Uppsala Symp. 1970, UNESCO.*

Beardsell, M. F., and Mitchell, K. J. (1973). Transpiration and
    photosynthesis in soybean. *J. Exp. Bot. 24*, 587-595.

Campbell, C. A., McBean, D. S., and Green, D. G. (1969). Influ-
    ence of moisture stress, relative humidity and oxygen diffu-
    sion rate on seed set and yield of wheat. *Can. J. Plant Sci.
    49*, 29-37.

Carlson, R. E., Yarger, D. N., and Shaw, R. H. (1972). Environ-
    mental influences on the leaf temperature of two soybean
    varieties grown under controlled irrigation. *Agron. J. 64*,
    224-229.

Demidenko, T. T., and Golle, V. P. (1939). Influence of relative
    air humidity on the yield and uptake of nutrient elements by
    the sunflower. *Comptes Rendus (Doklady) de l'Academie des
    Sciences de l'USSR 25*, 328-332.

Downs, R. J. (1975). "Controlled Environments for Plant
    Research." Columbia Univ. Press, New York.

Dunning, J. A., and Heck, W. W. (1977). Response of bean and
    tobacco to ozone: Effect of light intensity, temperature and
    relative humidity. *J. Air Pollut. Contr. Assoc. 27*,
    882-886.

Ford, M. A., and Thorne, G. N. (1974). Effects of atmospheric
    humidity on plant growth. *Ann. Bot. 38*, 441-452.

Forde, B. J., Mitchell, K. J., and Edge, E. A. (1977). Effect
    of temperature vapour-pressure deficit and irradiance on

transpiration rates of maize, paspalum, westerwolds, perennial ryegrasses, peas, white clover, and lucerne. *Aust. J. Plant Physiol. 4*, 889–899.

Fortanier, E. J. (1957). De beinvloeding van de bloei by *Arachis hypogaea* L. *Meded. Land. Wageningen 57*, 1–116.

Freeland, R. O. (1936). Effect of transpiration upon the absorption and distribution of mineral salts in plants. *Amer. J. Bot. 23*, 355–362.

Hoffman, G. J. (1973). Humidity effects on yield and water relations of nine crops. *Trans. ASAE 16*, 164–167.

Hoffman, G. J., and Jobes, J. A. (1978). Growth and water relations of cereal crops as influenced by salinity and relative humidity. *Agron. J. 70*, 765–769.

Hoffman, G. J., and Rawlins, S. L. (1970). Design and performance of sunlit climate chambers. *Trans. ASAE 13*, 656–660.

Hoffman, G. J., and Rawlins, S. L. (1971). Growth and water potential of root crops as influenced by salinity and relative humidity. *Agron. J. 63*, 877–880.

Hoffman, G. J., Rawlins, S. L., Garber, M. J., and Cullen, E. M. (1971). Water relations and growth of cotton as influenced by salinity and relative humidity. *Agron. J. 63*, 822–826.

Hughes, A. P. (1966). The importance of light compared with other factors affecting plant growth. *In* "Light as an Ecological Factor" (R. Bainbridge, E. C. Evans, and D. Rackhan, eds.), pp. 121–146. Wiley, New York.

Kinbacher, E. J. (1962). Effect of relative humidity on the high-temperature resistance of winter oats. *Crop Sci. 2*, 437–440.

Krizek, D. T., Bailey, W. A., and Klueter, H. H. (1971). Effects of relative humidity and type of container on the growth of F_1 hybrid annuals in controlled environments. *Amer. J. Bot. 58*, 544–551.

Lee, T. A., Ketring, D. L., and Powell, R. D. (1972). Flowering and growth response of peanut plants (*Arachis hypogaea* L.

var. Starr) at two levels of relative humidity. *Plant Physiol. 49,* 190-193.

Lunt, O. R., Oertli, J. J., and Kohl, H. C. (1960). Influence of environmental conditions on the salinity tolerance of several plant species, pp. 560-575. 7th Int. Congr. Soil Sci., Madison, Wisconsin.

McPherson, H. G., and Boyer, J. S. (1977). Regulation of grain yield by photosynthesis in maize subjected to a water deficiency. *Agron. J. 69,* 714-718.

Nieman, R. H., and Poulsen, L. L. (1967). Interactive effects of salinity and atmospheric humidity on the growth of bean and cotton plants. *Bot. Gaz. 128,* 69-73.

Nightingale, G. T., and Mitchell, J. W. (1934). Effects of humidity on metabolism in tomato and apple. *Plant Physiol. 9,* 217-236.

Nonnecke, I. L., Adedipe, N. O., and Ormrod, D. P. (1971). Temperature and humidity effects on the growth and yield of pea cultivars. *Can. J. Plant Sci. 51,* 479-484.

O'Gara, R. (1956). "Air Pollution Handbook." Section 9, 2, 4. McGraw-Hill Book Co., New York.

O'Leary, J. W. (1975). The effect of humidity on crop production. *In* "Physiological Aspects of Dryland Farming" (U. S. Gupa, ed.), pp. 261-280. Oxford and IBH Publ. Co., New Delhi, India.

O'Leary, J. W., and Knecht, G. N. (1971). The effect of relative humidity on growth, yield, and water consumption of bean plants. *J. Amer. Soc. Hort. Sci. 96,* 263-265.

O'Leary, J. W., and Knecht, G. N. (1972). Salt uptake in plants grown at constant high relative humidity. *Arizona Acad. Sci. 7,* 125-128.

Otto, H. W., and Daines, R. H. (1969). Plant injury by air pollutants: Influence of humidity on stomatal apertures and plant response to ozone. *Science 263,* 1209-1210.

Pareek, O. P., Sivanayagam, T., and Heydecker, W. (1969). Relative humidity: A major factor in crop plant growth, pp. 92–95. Univ. Nottingham School of Agriculture Rept. 1968–1969.

Prisco, J. T., and O'Leary, J. W. (1973). The effects of humidity and cytokinin on growth and water relations of salt-stressed bean plants. *Plant and Soil 39,* 263–276.

Rosenberg, N. J. (1974). "Micro-climate: The Biological Environment." Wiley, New York.

Sale, P. J. M. (1970). Growth and flowering of cacao under controlled atmospheric relative humidities. *J. Hort. Sci. 45,* 119–132.

Swalls, A. A., and O'Leary, J. W. (1975). The effect of relative humidity on growth, water consumption, and calcium uptake in tomato plants. *Arizona Acad. Sci. 10,* 87–89.

Tibbitts, T. W. (1978). Humidity. *In* "A Growth Chamber Manual: Environmental Control for Plants" (R. W. Langhans, ed.), pp. 57–79. Cornell Univ. Press, Ithaca, New York.

Tibbitts, T. W., and Bottenberg, G. (1976). Growth of lettuce under controlled humidity levels. *J. Amer. Soc. Hort. Sci. 101,* 70–73.

Tromp, J., and Oele, J. (1972). Shoot growth and mineral composition of leaves and fruits of apple as affected by relative air humidity. *Physiol. Plant 27,* 253–258.

van Marrewijk, N. P. A., and Visser, D. L. (1978). The effect of relative humidity on incompatibility and fertility in *Brassica oleracea* L. *Neth. J. Agr. Sci. 26,* 51–58.

Went, F. W. (1957). "The Experimental Control of Plant Growth." Chronica Botanica, Waltham, Massachusetts.

Wiersum, L. K. (1966). Calcium content of fruits and storage tissues in relation to the mode of water supply. *Acta. Bot. Neerl. 15,* 406–418.

Winneberger, J. H. (1958). Transpiration as a requirement for growth of land plants. *Physiol. Plant 11,* 56–61.

Woodward, R. G., and Begg, J. E. (1976). The effect of atmospheric humidity on the yield and quality of soya bean. *Aust. J. Agr. Res.* 27, 501-508.

Zimmerman, L. H. (1972). Effect of temperature and humidity stress during flowering on safflower (*Carthamus tinctorious L.*). *Crop Sci.* 12, 637-640.

HUMIDITY: CRITIQUE I

G. W. Thurtell

Soils Department, University of Guelph

Guelph, Ontario, Canada

The relationships between the water status of a plant and its environment are complicated and not completely understood. As noted by the keynote speaker, humidity levels have been observed to affect plant growth and development, rates of CO_2 exchange, flowering, nutrient transport, and air pollution susceptibility. However, these effects are generally the response of a plant to its total environment and are not uniquely related to the humidity of the air.

Water flows within the soil matrix to the surface of plant roots, enters the roots and root vascular system and moves within the conducting tissues of the xylem to leaves. This movement of water is caused by negative water potentials developed in cell walls at sites of evaporation in leaves and other transpiring tissue. It is also the interaction between this negative cell-wall water potential and the osmotic potentials within cells that determines tissue hydration. In general terms, soil water potential, soil water conductivity, root resistance (including the effects of root geometry and distribution), stem resistance, leaf resistance to water flow and transpiration rate, in addition to any osmotic, gravitational or temperature components of water potential, all interact to produce the water potential in cell walls and its effects on cell hydration. Each different set of

173

environmental conditions will have its effect on plants, and
plants grown in controlled environments often differ markedly
from field grown plants. We have found that cell osmotic potentials, in particular, differ between field and controlled-
environment grown plants. It is unlikely that these differences
result from unrealistic humidity conditions in controlled-
environment facilities because humidity conditions in chambers,
even though not under good control, do not have excessive devia-
tions from the humidity conditions in field environments. It is
more common for these differences to result from low levels of
photosynthetically active radiation, abnormal rooting volumes and
the rate and intensity of soil drying cycles. Both diurnal and
the longer-period cycles of tissue hydration level associated with
irrigation scheduling are important for development of normal
cell osmotic potentials and the transport of nutrients to non-
transpiring tissue.

The rate of transpiration from leaf tissue is proportional
to the difference between the vapor pressure in the leaf inter-
cellular spaces, adjacent to cell walls, and the vapor pressure
in the air external to the leaf. Transpiration rate is also
inversely proportional to the diffusive resistance of the system
which is normally described as the sum of stomatal and boundary
layer resistances. The vapor pressure in the air adjacent to the
cell walls (boundary layer) is a function of the leaf temperature
and the other components of the leaf water potential, and is
often very different from the vapor pressure of the air at a
distance from the leaf. Boundary layer resistances are dependent
on leaf size, shape and orientation and on ventilation rate.
Stomatal resistances are dependent on irradiance levels, tissue
water status and CO_2 levels. While the boundary layer and
stomatal resistances quantitatively describe the diffusive
exchange process, they are sometimes difficult to determine
adequately and often are ignored even though they usually differ
markedly between natural field and controlled environments. When

resistances are required for the calculation of experimental variables, e.g. the calculation of dosages in air pollution studies, the results obtained by different researchers cannot be compared if these resistance measurements have not been made.

It is clear that the water status of plants is complicated and is influenced by many environmental parameters. The humidity of the air is one of the most important of these parameters and it should always be reported. Dew point, mixing ratio, or relative humidity, plus the temperature at the measurement site, are all suitable units for reporting. Relative humidity is commonly chosen when relative humidity sensors are employed in the control system.

While psychrometers are commonly used sensors for both humidity recording and control, a number of other sensors are also in common usage for these purposes. The lithium chloride dew cell, in which the temperature of a saturated solution of lithium chloride is heated until its vapor pressure is equal to that of the surrounding air, is frequently used for humidity measurements. Sulphonated polystyrene, "Dunmore" and "Vaisala" sensors provide measurements of relative humidity and are used for both the recording and controlling of humidity in controlled environment facilities. While all of these sensors have been used success-fully, all must be maintained carefully and it is most important that they be mounted where they are readily accessible for cleaning and inspection. It is common practice to use different sensors for control and record functions. This provides a continuous check on the behavior of the control system. Contin-uous recordings of both temperature and humidity are a most desirable feature in many controlled-environment facilities.

HUMIDITY: CRITIQUE II

John S. Forrester

Scientific Systems Corporation
Baton Rouge, Louisiana

INTRODUCTION

Controlled environments designed to meet specified performance levels require critical evaluation of many factors. These
factors must be considered not only independently, but even more
importantly in relation to each other and as a part of the final
system.

This systems-oriented approach is particularly important in
the design of humidity controlled environments intended for plant
growth applications. Once the capacity and performance specifications have been selected or "fixed", so are items previously
considered as variables: plant load, temperature range, humidity
range, radiation level, and air velocity through the plant canopy.
The experimental requirements and needs of the plants have
already determined which variables are no longer available for
the subsequent system design process. Variables that do remain
include types of controllers and sensors, coil geometry and air
flow in the conditioning system, thermal (heating and cooling)
design and capacity, and subsystems related to water addition or
removal. While the characteristics of each of these variables
when examined individually are of interest, their interaction

177

will determine the performance level of the controlled plant
growth environment.

HUMIDITY SENSORS

   The characteristics of various humidity sensors have been
reviewed by Wiederhold (1978). Unfortunately, the ideal or
universal water vapor sensor has not yet been developed, and so
selection of a sensor of a type particularly suitable for a
given design application becomes a matter of compromise. The
speed of response, need for maintenance, precision, accuracy, cost,
and overall performance under the intended design conditions
are factors that must enter into the selection process. Fig. 1
illustrates the useful operating ranges for different moisture
sensors. Three of these sensors are of a type most frequently
selected for plant growth systems: psychrometer, saturated salt
(LiCl) dew point sensor, and the Pope cell electrical resistance
sensor.

   While the psychrometer type sensor has been widely used and
is theoretically attractive for use under conditions found in
plant growth environments, it has been difficult to maintain and
inconvenient to use. Most frequent complaints include wick
contamination due to airborne dirt or spores, and the need to
translate "wet bulb temperature" to relative humidity. With
the help of suitable psychrometric tables the advent of microprocessors now makes it possible to obtain the readout, if
desired, in terms of relative humidity, but the contamination
problems remain.

   The lithium chloride dew point sensor has been very useful in
applications requiring good sensor performance above and below
the comfort range, i.e. at dew points below 5° and above 35°C.
Since dewpoint represents an absolute moisture measurement, the dew point sensor is proving especially attractive for

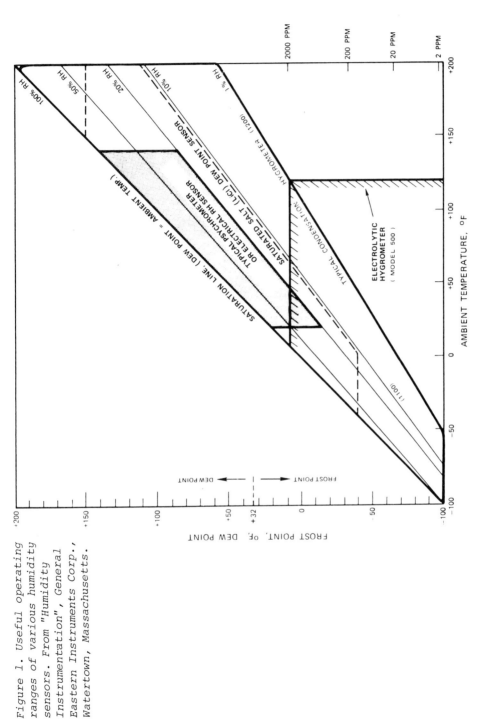

Figure 1. Useful operating ranges of various humidity sensors. From "Humidity Instrumentation", General Eastern Instruments Corp., Watertown, Massachusetts.

plant growth related applications. The sensor consists of an absorbent fabric bobbin covered with a bifilar winding of inert electrodes and coated with lithium chloride. Good design practices dictate that the sensor power supply be self-limiting and, if possible, retain power to the sensor during times when other power sources may be shut down.

The Pope cell electrical relative humidity sensor is a reasonable compromise with regard to both cost and performance for use in plant growth systems. The sensor element is a bifilar conductive grid deposited on a sulfonated polystyrene ion exchange substrate. The basic sensor has three characteristics which, if ignored, will result in control and readout devices unacceptable for use in high performance systems: temperature sensitivity, non-linear (logarithmic) response, and slight hysteresis. The non-compensated sensor is, at best, of marginal use at fixed temperatures and in the 35 to 80% relative humidity range. By electronically linearizing and temperature compensating the sensor output, substantially improved performance can be obtained over the entire temperature and relative humidity range of interest in plant growth applications. A temperature and relative humidity recorder based on this approach is shown in Fig. 2. Note also the spectral aluminum (Alzak) sensor housing with built in aspirator motor, a design which safeguards against potentially false readings from high radiation loads as might be encountered in plant growth environments.

One rather unique approach of circumventing shortcomings associated with moisture sensors is to control moisture within the chamber by equilibrating the circulating air with water at the desired dew point temperature. Thus the dew point within the chamber will be the temperature of the equilibrated air which can be measured and controlled with a standard temperature controller. This air is then reheated to the desired dry bulb temperature before reintroduction into the chamber. While

FIGURE 2. Temperature and relative humidity recorder.

limited in application, this approach has been found very useful
in the design of some high performance, small growth cabinets.

A very much neglected requirement has been the need to
calibrate humidity sensors in the field. The assumption is
generally made that suitable calibration techniques are at best
difficult and time consuming. With the availability of a
recently introduced humidity sensor calibration kit[1], such
calibration should become standard procedure for users of
humidity controlled systems. The commercially available kit is
based on ASTM Standard 104-51 method C, and provides premixed
saturated salt mixtures covering the range of 34 to 90% relative
humidity along with the accessories necessary for simple sensor
calibration.

[1]*Abbeon Cal. Inc. Gray Avenue, Santa Barbara, California   93101.
Catalogue No. 263. Kontrol (Sic) Hygrometer.*

CONTROLLERS

    In selecting a suitable humidity (or temperature) controller,
it must again be remembered that this controller by itself does
not control the final performance level of the end product, e.g.
the degree of humidity control in the plant growing area.
Frequently the tendency is to blame the controller for performance shortcomings actually originating in other parts of the
controlled environment system. In most humidity controlled
installations examined by the author, the controller was actually
the strongest link in the entire control system chain.

    Regardless of the design, the controller only provides some
type of initiating signal: an on-off contact closure, a time
proportioning series of pulses, or analog signals based on
current, voltage, or control air pressure and varying in relation
to controller demand. What happens then? While on-off signals
can be used to open and close refrigerant lines, turn heaters
on and off, control water and steam solenoids, or turn blowers
on or off, this type of logic may not suffice for humidity
control requirements. Frequently the need exists to modulate
heating and cooling devices, and it is in this application that
proportioning controllers are of primary interest. While
proportioning electric heating is easily instrumented, proportional cooling is not. There has been continued need for a
commercially available, low capacity modulating valve suitable
for controlling refrigerant through small diameter lines. In
the absence of a satisfactory device of this type, equipment
designers have had to find their own solution when faced with
the problem of how best to interface proportional controller
output with a refrigerant or secondary coolant system.

    With the advent of microprocessor based systems, control
technology has taken on a new dimension. The first commercially
available units incorporated only set point, programming and
display features, with newer units including temperature, humidity

and lighting control and data processing. Even with these
advances, use of the microprocessor is analogous to an elephant
pulling a garden cart. The full utilization of computer
technology requires valid mathematical models covering the key
variables of the controlled system. With such information, the
microprocessor can be used to effect optimum control along with
another increasingly important requirement, i.e. minimum energy
consumption. As shown later, substantial progress has been made
in this direction.

HEAT LOADS

   Most important to the design of environmentally controlled
plant growth installations is the full consideration of all
contributory sensible and latent heat loads. Major heat sources
include the radiation load from lamps, as well as heat from
condensation of water vapor and mechanical sources such as fans
or blowers.

   Heat loads originating from internally located air movers
are becoming even more important as the need for higher air flow
through the plant canopy is recognized (van Bavel, 1973). Greater
air flow requirements mandate larger motors, which in turn
impose greater heat loads on the system. Rather than making
proper allowance for this fact, designers frequently are too
concerned with room wall and floor losses; with plant growth
installations operating around ambient conditions, wall and floor
losses (or gains) may be almost negligible.

   Plant transpiration loads and resulting condensation of
water on cooling coils have frequently been the stepchild of
heat load calculations. Such loads are often all but ignored in
the design of plant growth chambers. As a result, many existing
installations are plagued by inadequate water removal capacity,
leading to an inability to meet operating conditions that require

low dew points under moderate or heavy plant loads. It should
also be remembered that demands for moisture removal are unusually
high with certain plants, e.g. legumes, and under high lighting
intensities.

Table 1 summarizes typical heat loads that can be expected
in a large plant growth room when operating around ambient
conditions with moderate illumination levels and high transpiration loads. Only when all the thermal loads have been
calculated and determined in this manner, can the conditioning
system be designed.

THE CONDITIONING SYSTEM

The conditioning system can be considered the key to the
entire controlled environment system and thereby that portion of
the plant growth chamber primarily responsible for setting the
level of performance for humidity control. The conditioning
system includes not only the coils, blowers, humidifiers, heaters
and related components but, most important, incorporates the
sizing, geometry, and overall combination of all of these as
necessary to achieve a desired degree of performance and control.

Once the size of the growth area has been determined and the
desired air flow through the plant canopy has been specified,
the total amount of air to be conditioned becomes another "fixed"
quantity. For the example shown in Table 1, the conditioning
system blowers and coils must handle 270 m^3 (9,600 ft^3) per
minute. This requirement, along with the thermal load calculations covering the desired operating parameters, then become
the starting points for the design of the remaining conditioning
system.

The design process must take into account three major areas:
the primary heating and cooling system, coil and air flow
geometry, and most important yet probably least recognized, the

TABLE 1. Summary of Sensible and Latent Heat Loads for a
Plant Growth Room

Design Conditions

Plant bed	6 m^2 (64 ft^2)
Dry bulb temperature	25°C
Dew point temperature	17°C (43% RH)
Lamp bank (with plexiglas barrier)	44 VHO (very high output) 8' fluorescent lamps and 24 ea 100 W incandescent lamps
Air flow	46 m min^{-1} (150 ft min^{-1})
Make-up air	2.8 m^3 min^{-1} (100 ft^3 min^{-1})

Thermal Loads

	kg cal hr^{-1}	Btu hr^{-1}
Surface losses	nil	nil
Blower motor	2855	11,330
Radiation load (approx)	5095	20,220
Transpiration load	4101	16,275
	(20)	(80)
Total system load	12,301	47,745

heat and mass transfer characteristics of the selected system.
A plant growth environment is a dynamic system. Lighting levels
change, transpiration rates vary, temperatures and moisture levels
rise and fall throughout the plant growth cycle. Only when the
conditioning system is properly designed to effectively respond
to such changes, can the plant growth system perform properly.

To illustrate the importance of properly recognizing and
allowing for heat and mass transfer characteristics, four
different conditioning systems were examined. In all cases the
assumption was made that the control system was proportioning and
coil size was not a limiting factor. The results are shown in
Figures 3 to 6.

The majority of commercial growth rooms and cabinets
currently in use incorporate a single conditioning coil, without
by-pass, i.e. all of the air stream passes through the coil.
Fig. 3   illustrates this case. With only a temperature controller, humidity levels will vary as the water load changes.
At the same time the humidity level can be shifted by addition
(or subtraction) of sensible heat, e.g. with an electric heater.
As the sensible heat to air ratio decreases, the curves tend to
become almost flat, suggesting limited application of this system
for maintaining elevated humidity levels without the need for a
humidity controller.

Fig. 4   illustrates the case of the same single coil, without
by-pass system, when instrumented with both temperature and
humidity controllers, a humidifier and a heater. Note that five
variables influence the performance of this system: (1) sensible
heat, including heaters, (2) air circulation, (3) water input
(from transpiration and humidifier), (4) temperature, and (5) the
absolute humidity operating points. Every one of these "variables",
with the exception of the heater and humidifier inputs, is
actually "fixed" by virtue of the performance specifications; the
ratio of the heater to humidifier input must be adjusted if

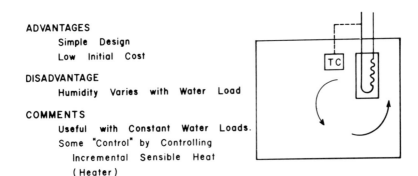

ADVANTAGES
    Simple Design
    Low Initial Cost

DISADVANTAGE
    Humidity Varies with Water Load

COMMENTS
    Useful with Constant Water Loads.
    Some "Control" by Controlling
        Incremental Sensible Heat
        (Heater)

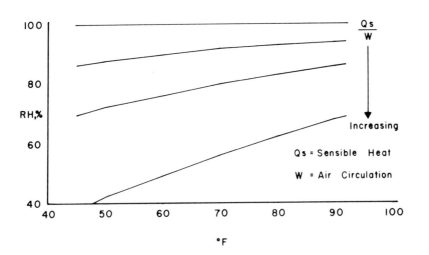

FIGURE 3. Single coil conditioning system with temperature control only.

optimum operation is to be achieved. It is evident that current systems rarely meet this requirement, with a resulting, detrimental effect on humidity control.

Fig. 5   covers a modification of the single coil system in that a second coil has been added. This approach has been widely used in plant growth rooms. The performance of such an installation can be projected by determining the operating parameters of each coil independently. Response to humidity change tends to be slow, energy requirements moderately high, and refrigeration circuiting more complex than for a single coil approach.

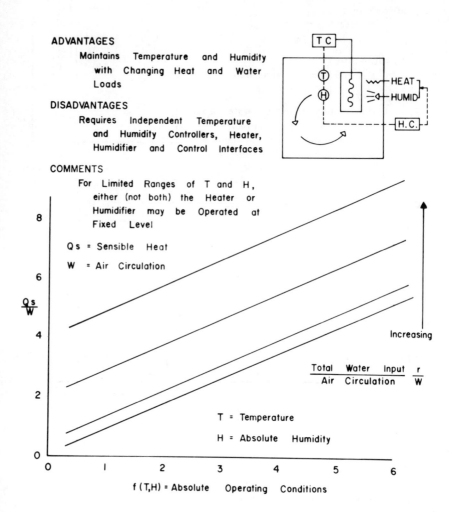

ADVANTAGES

Maintains   Temperature   and   Humidity
with   Changing   Heat   and   Water
Loads

DISADVANTAGES

Requires   Independent   Temperature
and   Humidity   Controllers,   Heater,
Humidifier   and   Control   Interfaces

COMMENTS

For   Limited   Ranges   of   T   and   H,
either   (not   both)   the   Heater   or
Humidifier   may   be   Operated   at
Fixed   Level

Qs   =   Sensible   Heat

W   =   Air   Circulation

$\dfrac{Qs}{W}$

Increasing

$\dfrac{\text{Total   Water   Input}}{\text{Air   Circulation}}$ $\dfrac{r}{W}$

T   =   Temperature

H   =   Absolute   Humidity

f (T,H) = Absolute   Operating   Conditions

FIGURE 4.   Single coil conditioning system with temperature and humidity control.

ADVANTAGES

    Relatively   Simple   Control   System

    Operating   Parameters   of   each   Coil   Independently
      Determined

DISADVANTAGES

    May   be   Capacity   Limited

    More   Complex   Refrigeration   than   Single   Coil   System

    Slow   Dehumidification   Response

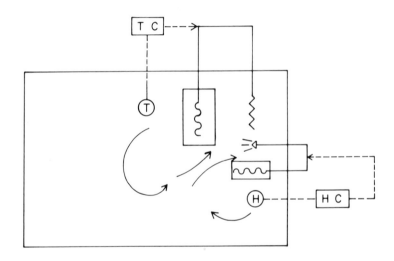

FIGURE 5.   Two coil conditioning system with temperature and humidity control.

If air is passed over and equilibrated with the coil surface
at a controlled reduced temperature, this becomes a very exact
means of setting the dew point temperature. Reheating the air
to the operating dry bulb temperature then establishes both of
these parameters. For controlled plant growth environments, this
direct approach quickly becomes extremely energy inefficient;
a small reach-in cabinet can require 10 HP of refrigeration and
an equally unacceptable amount of electric reheat. On the other
hand, if a portion of the circulating air stream is bypassed,
and the amount of this bypass can be regulated, the result is a
growth chamber with straight line temperature and humidity control,
excellent performance over a wide range of dew points, and minimal
energy consumption. This design was first reported by van Bavel
(1970), The first commercial version incorporated manual bypass
control, and later models added a number of refinements. Fig. 6
illustrates the design and characteristics of a one-coil air
handler incorporating programmed bypass, temperature and humidity
control. A microprocessor provides not only the temperature and
humidity controller functions, but also controls the ratio of the
controlled to bypassed air. In this approach, the requirements
for reheat and humidification are minimal, making the chamber
very energy efficient.

HUMIDIFICATION AND DEHUMIDIFICATION

It should be evident at this point that, depending on the
design of the conditioning system, demands placed on humidifiers
may range from minimal to severe. Failure to properly select and
design the conditioning coil geometry and air flow may impose
unnecessarily severe requirements on the accessory humidifying
and dehumidifying devices; in extreme cases satisfactory compensation for deficient design may not even be possible. The
existence of such shortcomings may give the impression that a

ADVANTAGES
    Excellent,   Straight   Line   Control
    Energy   Efficient,   Minimal   Reheat
    Fast   Response
    Simple   Refrigeration   and   Humidification

DISADVANTAGES
    Requires   More   Instrumentation
    or   Computer   Model   and   Microprocessor

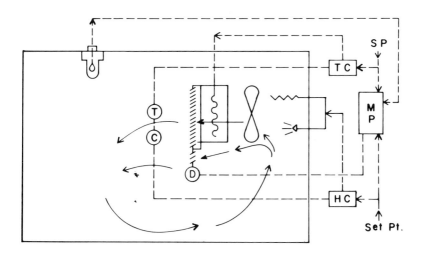

FIGURE 6. *Single coil conditioning system with temperature and humidity controls and programmed by-pass.*

particular chemical drier or humidifier was improperly sized, when in fact the problem was caused by improper design of the overall conditioning system. With the plant load itself constantly adding to the circulating water concentration, the answer to good control at high humidities is not found with a humidifier of greater capacity, but by proper design of the conditioning system.

There are other factors, less well recognized, that can impair equipment performance. When operating at dew point temperature less than one or two degrees below the dry bulb temperature, an absolutely flat temperature profile across the coil becomes mandatory. Cold spots along the coil and associated tubing can

lead to a serious loss of control. This is one reason why
excellent high dew point performance has frequently been achieved
with secondary cooling systems, i.e. chillers. The cooling coil,
in combination with the precisely temperature controlled chiller
brine, provides a constant heat sink. In contrast, a DX (direct
expansion) refrigerant coil with absolute temperature uniformity
over a wide range of load conditions is difficult to achieve.

    With properly designed systems, there should not be any need
for chemical driers except to bring about dew points below 4
degrees. When chemical drier systems are selected, it must be
remembered that the drying agent has a characteristic equilibrium
curve, and the effect on the incoming and outgoing air streams
must be properly calculated. Most important, the refrigeration
system must effect dehumidification to a minimum dew point, so
that the chemical drier is presented the least possible moisture
input. With this approach, a silica gel drier should never need
to process incoming air at the moisture levels exceeding 40 grains
per pound of dry air, providing output moisture levels below 10
grains per pound at all operating temperatures up to 25°C. The
most popular commercial drier used for growth chambers is a
4250 liters per minute model. Assuming the worst case, i.e. a
35°C operating point, this drier will remove approximately 1.2 kg
of water per hour. The designer must decide whether or not this
meets the physiological characteristics of the plant load. If
not, chemical driers with larger capacity are readily available.

REFERENCES

van Bavel, C. H. M. (1973). Towards realistic simulation of the
    natural plant climate. *In* "Plant Response to Climatic
    Factors" (R. O. Slatyer, ed.), pp. 441-446. Proc. Uppsala
    Symp. 1970, UNESCO.
Wiederhold, P. R. (1978). Which humidity sensor? Inst. Contr.
    Systems 51, 31-35.

HUMIDITY: GUIDELINES

L. A. *Spomer*

Department of Horticulture
University of Illinois
Urbana, Illinois

Water is the most common liquid on earth and as such exerts
an overwhelming influence on plant growth and survival. Plants
contain and use large quantities of water. This water is more
than merely an inert filler; probably every plant growth process
is directly or indirectly influenced by the status of plant water
which depends primarily on the balance between water absorption
from the soil and water loss to the atmosphere. Because of the
obvious effect of a lack of water to plant roots, investigators
usually monitor the environmental factors affecting water availability. However, researchers tend to largely ignore the factors
affecting water loss, the most important of which probably is
humidity.

Although humidity represents only about 0.001% of the earth's
total water supply, it is the atmosphere's third most abundant
component and is an important factor influencing plant growth.
Natural atmospheric water vapor concentrations average 3% by mass
but vary from nearly zero to about 4%. A similar range of water
vapor concentrations in found in controlled environments; however, the average is typically higher. Humidity in controlled
environments is characterized by large fluctuations as a result
of opening of chambers, irrigating plants, and condensation on

193

heat exchangers. Therefore, it is especially advisable to monitor humidity in controlled environment facilities in order to
ensure full and proper interpretation of experimental data. This
paper briefly proposes guidelines for doing so. Emphasis is
placed on defining the proper units of measurement.

TERMS AND UNITS

The term *moisture* technically refers to the water content of
any solid, liquid, or gaseous material. However, *moisture* is
customarily used to describe only the liquid water content of a
solid or liquid whereas *humidity* seems to be the preferred term
to describe the water *vapor* content of a gas. It is proposed
that either the term *water vapor* or *humidity* be used to depict
water vapor in the air.

Several different terms are commonly used to describe humidity, some of which are not directly related to the effects of
humidity on plant processes. The most significant of these processes is transpiration or water vapor flux (E) from the leaf.
Transpiration directly affects plant water status and heat balance. The "driving force" for transpiration is the difference
between the water vapor concentration at the leaf evaporating
surface (ρ_{vs}) and that of the surrounding atmosphere (ρ_{va}):

$$E = (\rho_{vs} - \rho_{va})r_v^{-1} \qquad (= g\ m^{-2}\ s^{-1}),$$

where r_v is the vapor diffusion resistance between the leaf evaporating surface and the surrounding atmosphere which depends on
leaf surface properties and turbulence in the surrounding atmosphere. These direct and indirect effects of humidity are all
related to the *water vapor concentration* of the air, also referred to as *vapor density* or *absolute humidity* ($g\ m^{-3}$ or $mol\ m^{-3}$).

Water vapor concentration can be expressed in several terms
which, for the most part, can be derived from one another through
the use of appropriate tables or charts. Possibly the most familiar such term is *relative humidity* which is the ratio of ambient vapor concentration (ρ_v) to that at saturation (ρ_v') at the
same air temperature:

$$h_r = \rho_v \, \rho_v'^{-1} \qquad\qquad (X\ 100 = \%).$$

It should be noted that relative humidity changes with temperature even though water vapor density does not. This has led to
considerable confusion and error in past interpretation of humidity effects on plant growth and therefore is probably not the
best unit to describe humidity. *Saturation vapor density* of air
is that which occurs at equilibrium with a free water surface at
the specified temperature. *Specific humidity*, another expression
of water vapor concentration, is the mass concentration of water
vapor (h_s) in a given mass of moist air (m_v') where dry air mass
is m_a: and water vapor mass is m_v.

$$h_s = m_v \, m_v'^{-1} = m_v (m_v + m_a)^{-1} \qquad (= g\ g^{-1}),$$

An additional measurement, atmospheric pressure is required to
convert specific humidity to vapor density. Specific humidity
and *mixing ratio* (r) are commonly used by heating and air conditioning engineers, but seem of limited use for plant scientists.
The mixing ratio is the ratio of the mass of water vapor to that
of the dry air with which it is associated:

$$r = m_v \, m_a^{-1} \qquad\qquad (= g\ g^{-1}),$$

Dew point temperature, the temperature at which air just satu-
rates when cooled without changing its water content or the barometric pressure, and *wet bulb temperature,* the temperature of a
wetted thermometer exposed to an unsaturated atmosphere, are two
additional indicators of water vapor density. Wet bulb temperature is the simplest and therefore probably the most common

method of measuring humidity, either directly or for derivation
of other units. Air temperatures are required to convert either
dew point or wet bulb temperatures to vapor density. *Vapor saturation deficit,* the difference between ambient and saturation vapor densities:

$$VSD = \rho_v - \rho_v' \qquad\qquad (= g\ m^{-3}),$$

is probably a more accurate descriptor of actual evaporation conditions than the others because it is a measure of the magnitude
of the driving force for transpiration.

Water vapor pressure (e) is often used to describe humidity.
It is related to vapor density by the perfect gas law:

$$e = 4.62\ x\ 10^{-4}\ x\ \rho_v\ x\ T \qquad\qquad (= kPa,\ T = {}^{\circ}K).$$

Relative humidity defined in terms of vapor pressure is the ratio
of ambient vapor pressure (e) to saturation vapor pressure (e'):

$$h_r = e\ e'^{-1} \qquad\qquad (x\ 100\ =\ \%)$$

Likewise, vapor saturation deficit is the difference between ambient and saturation vapor pressures:

$$VSD = e - e' \qquad\qquad (= kPa).$$

In conclusion, humidity in controlled environments is probably best described in terms of water vapor density ($g\ m^{-3}$) or
other units from which water vapor density can be readily derived
(ie. relative humidity, vapor pressure, dew point temperature,
wet bulb temperature, vapor saturation deficit); however, simul-
taneous air temperature measurement must accompany all of these.
Vapor saturation deficit is possibly the most useful unit relative to transpiration (see paper by Hoffman in this volume).

HOW TO MEASURE

   Ideally, humidity measurement should be made with the most
direct or primary sensor possible to a precision equal to or
exceeding the precision resolvable by plant response. The actual
choice of measuring instrument depends on many factors including
accuracy and precision, need for remote measurement (chamber
closed), speed (response time), difficulty, and availability.
Instruments for direct measurements include psychrometers, dew
point temperature sensors, and infrared gas analyzers. Several
indirect methods utilizing sensor expansion, resistance, capacitance, etc. are also frequently used for control and continuous
recording but have not been as useful for precision measurement.
It is recommended that direct humidity measurements be made regularly and used for calibration or at least used to frequently
calibrate indirect measurement sensors.

WHEN TO MEASURE

   Continuous monitoring of humidity, where practicable, is recommended. A minimum recommended measurement is once during each
different daily temperature and light regime for a period sufficient to obtain an accurate measurement relative to chamber
steady state heating and cooling cycles and instrument response
time.

WHERE TO MEASURE

   Sampling should be done in a location representing the average conditions to which the plants are exposed. Although humidity gradients commonly occur within growth chambers, they are

usually of little significance. The sampling location for humidity is therefore probably much less critical than that for light
or temperature. It is recommended that humidity be measured at
the top of the plant in the center of the growing area. Air
temperature measurements for humidity derivations must be made at
the same location.

WHAT TO REPORT

The minimum recommended information reported should include
the average and extreme of reading for each temperature and light
level. These measurements should be expressed in the units in
which they were measured and should be accompanied by simulta-
neous air temperature measurements.

In conclusion, these guidelines suggest minimum recommendations for humidity measurement. Specific measurement practices
should be adapted to provide understandable characterization of
humidity in each specific situation. It is strongly recommended
that a uniform terminology relevant to plant processes should be
adopted.

HUMIDITY: DISCUSSION

HOFFMAN: Atmospheric moisture should be changed to something
more descriptive. Why specify methods of measurement for humidity
only and not other factors? Investigators should report saturation deficit and wet bulb temperature in specified units. Air
temperature should be measured at the same point as wet bulb temperature since both are required to describe vapor pressure.

McCREE: Absolute humidity should be reported rather than relative humidity.

HAMMER: Relative humidity should be specified because it is
probably still the most commonly used and most widely understood
measurement.

CURRY: Absolute or specific humidity should be specified
rather than relative humidity. Expressing humidity as $g \ g^{-1}$ means
something to me without further calculations.

KAUFMANN: Absolute humidity is $g \ m^{-3}$ and specific humidity is
$g \ g^{-1}$. Stomata respond to the water vapor difference between the
leaf and air so humidity should be expressed in these terms as a
saturation deficit. Frankly, I would not use any of the proposed
methods of measurement in the guidelines. Some of the indirect
methods work well enough to describe plant response to humidity if
checked periodically for accuracy. Plants do not have the capacity
to resolve fine differences in humidity. I suggest keeping humidity measurements simple.

BATES: For consistency of units relative humidity would better
be described as percent, dew point temperature as ^{o}C, and specific
humidity as $g \ g^{-1}$. The unit $g \ g^{-1}$ could also be used to describe

mixing ratio and should be specified as to which quantity it
refers.

KRIZEK: Why aren't more growth chamber controls specified in
dew point rather than relative humidity as they are in Germany?
It should be recognized that tremendous fluctuations in humidity
occur as chambers are opened for working with the plants; this
can significantly affect pressure bomb readings if the plants are
taken out of the chamber for measurement. Many growth chambers
have virtual showers as a result of large droplets from humidifiers. I would like to see better humidity control methods employed
by manufacturers.

SEARLS: This is a good point. Dew point controllers are often used to control humidity, but we do what the customer wants.
If he requests relative humidity control, that is what he gets.
If he asks for dew point control, then that is what he gets.
Methods of humidification have improved by introducing humidity
into the air handling units separate from the chambers, mostly as
steam, so except for small chambers without air handling units,
this situation has improved.

FORRESTER: Agreed! The customer gets what he requests. Historically, relative humidity controllers were the first type used.
Investigators will have to pay more for better control. Some
dew point controllers are not very good. Water droplets in growth
chambers are technically not necessary, even when wide humidity
control ranges are utilized.

CAMPBELL: Investigators should specify humidity as concentration, dew point, or vapor pressure. It doesn't matter very much
which is used because all are easily converted to one another.
Thurtell mentioned that all of these measurements are rather con-
servative; that they will not vary much from day to night. However, relative humidity will likely vary significantly if there
are temperature changes from day to night. Which moisture condition is controlled or measured does not matter very much but in

any case, a concentration or vapor pressure should be reported because this is what the plant "sees";. The plant does not "see" relative humidity.

TIBBITTS: Isn't it the vapor pressure deficit which the plant actually "sees" and this varies with temperature and not the vapor pressure? It seems that we really should be looking at vapor pressure deficit.

CAMPBELL: But it really isn't the vapor pressure deficit that the plant "sees"; it is the difference between the saturation vapor concentration at the leaf and that of the surrounding air. With the information which Salisbury presented previously investigators can hopefully calculate leaf temperatures. But, if atmospheric water vapor concentrations have not been reported, the investigator is out of luck. Hence the vapor pressure or vapor concentration of the atmosphere is the most important single environmental measure relative to humidity. The other parts can be derived from energy balance calculations.

RAWLINS: Investigators should not report vapor pressure deficits without also reporting leaf temperature because that would not be meaningful.

TANNER: We can interconvert between units if we measure the dry bulb temperature plus the atmospheric vapor pressure, dew-point temperature, or relative humidity. If the investigator wants specific humidity, he also needs barometric pressure.

KLUETER: In response to Kaufmann's comments indicating that plants are unable to resolve small temperature or humidity changes; it all depends upon where plants are within the moisture range of air. In mid-range, this is probably true; but near either extreme, small changes may become critical for plants.

HAMMER: I recommend specifying relative humidity and air temperature because the equipment for measuring them is available to most investigators working in controlled environments.

McFARLANE: Why not express humidity in terms of the SI units $mol\ m^{-3}$?

KAUFMANN: Doing so may be too much of a change from that with which we are familiar.

KOLLER: I would recommend that humidity be measured as the average over the plant growing area rather than only at the center of the growing area.

CAMPBELL: We should restrict ourselves to a primary humidity measurement such as dew point temperature or wet bulb temperature, measurements that do not require continuous recalibration. Other methods do need continuous recalibration to maintain accuracy.

FORRESTER: Calibration kits for humidity sensors are commercially available. Some of the old sensors such as hair or nylon fiber are accurate to within 2% and should not be discounted.

TIBBITTS: Our experience indicates that these hair or nylon fiber devices do not maintain calibration and were therefore not recommended in the guidelines.

FORRESTER: I am referring to the sealed type sensors that are protected from direct exposure to the atmosphere and dust. They do exhibit some hysteresis, however.

TANNER: Crushed hair fiber sensors can maintain fairly good calibration. The sulfonated polystyrene sensors also exhibit hysteresis and have an undesirable feature in their very slow approach to the final reading. Again, regardless of the type of controller, assuming it provides adequate control, the main problem is how to accurately measure the humidity. For accurate measurements the dew point or psychrometric techniques are hard to beat.

TIBBITTS: Actually 3 separate humidity sensors are usually required in controlled chambers; one for precise experimental measurement, one for continuous recording, and one for control. Often requirements dictate that three different types of sensors be used for these three humidity related functions.

KAUFMANN: I do not agree with Thurtell that humidity conditions in growth chambers can be easily established to duplicate

humidity conditions in natural environments. For example, it is
very difficult to duplicate in growth chambers the low daytime
humidity levels of the arid western United States.

CAMPBELL: There is also a need to maintain night conditions
in growth chambers that are comparable to those in the natural
environment. Night conditions in the natural environment commonly
result in saturation deficits that are near zero. Saturation deficits in growth chambers at night are often much larger than this.
With large saturation deficits plant water potentials do not recover to near zero values during dark period, as is commonly observed outdoors during the night.

KAUFMANN: I would like to reinforce Hoffman's proposal to control atmospheric moisture by saturation deficits, or vapor pressure
differences between the leaf and air rather than on the basis of
vapor pressure of the air. We found that the temperature optimum
for photosynthesis was $7^{o}C$ higher when a constant vapor pressure
difference from leaf to air was maintained over a range of temperatures than when a constant vapor pressure only was maintained.

WENT: Our experiments showed that under ideal water supply
conditions there was practically no effect of different relative
humidities on plant growth. Hoffman, in your studies, were the
root conditions, and thus water supply conditions, less than ideal
to cause the indicated effects on growth at different atmospheric
humidity levels?

HOFFMAN: Our experiments at the Salinity Laboratory were performed in water culture so we had no water deficit conditions.
Your experiments were conducted under low light intensities while
our research was under sunlight conditions. Therefore the stomata
of your plants may not have been as responsive as stomata of our
plants.

WENT: I do not think that liquid culture eliminates water
stress. In fact, it may result in very poor water supply for
plants.

HOFFMAN: We used gravel culture and irrigated 24 times a day.
The pots were filled with water and allowed to drain so aeration
was not a problem. Krizek indicated that ornamental plants in
peat-vermiculite soil mix and watered 6 times a day also exhibited
responses to humidity. Thus I do not believe that water availability has been a limiting factor in most studies reporting humidity
effects upon plant growth.

WENT: There are enormous differences in root systems that develop with variations of frequency of watering. With frequent
watering, plants do not develop root hairs and this greatly influences the absorption of water.

TIBBITTS: I have not found measurement of humidity in chambers
to be easy. Most chambers are subject to rather large fluctuations
in relative humidity with the regular heating and cooling cycles.
This means that several measurements must be taken and averaged
to obtain representative readings. Measurement is complicated by
the fact that wet and dry bulb sensing elements, being of different mass, respond with different time constants to temperature
fluctuations. Thus considerable expertise is required to obtain
accurate readings.

THURTELL: My comments were directed more toward physical aspects of measurement. Instruments are available for accurate
measurement. There are, of course certain operational problems
that must be recognized.

ANDERSON: The operational problems are a major consideration.
When one has 40 or 50 chambers to maintain this is significant.

RULE: There are fewer problems in measuring humidity in many
chambers today than there were 10-15 years ago because many chambers have modulating systems, rather than on-off systems.

KAUFMANN: Some of us are using the Vaisala humidity units in
porometers. Are there individuals who can comment on the usefulness of these sensors for control or measurement of humidity in
growth chambers?

PAIGE: We fitted three growth chambers with Vaisala units for humidity control 1-1/2 years ago. They were placed in an aspirated chamber below the plant bed. The humidity conditions have been monitored and calibrated with an EG&G dew point hygrometer. We had problems only when we used the chambers for ozone fumigation. Then, there were some anomolous shifts in calibration that developed over periods of a few hours. Otherwise the control has been very stable over the 1-1/2 year period of operation of these chambers.

TIBBITTS: Can someone provide an explanation and description of these units?

KAUFMANN: The units are made in Finland, (Vaisala Oy, Helsinki 42, Finland). The sensor is a small capacitor chip with overlapping fingers of metal over the chip. The capacitor is part of an oscillating circuit in a small handle unit (6" long and 3/4" in diameter) powered by 3.6 VDC and providing an output of 0-100 mv for 0 to 100% relative humidity. The response is very fast, responding to ≈90% of a change in humidity in one second with little hysteresis, especially at humidities below 80%. The unit costs about $550.00 and is available from Weather Measure Company (HM 111P Solid state relative humidity sensor) for recording and controlling in chambers.

RULE: We decided against using Vaisala units because of the added cost to the consumer and a high failure rate of the units. However I believe the sensors are now coated and their failure rate may be lower.

There is a new control and read-out system for humidity produced by EG&G Inc. (Model 911) that is being used but it also is of rather high cost.

WALKER: Vaisala units are available in both coated and uncoated types. A replacement chip costs only about $125.00 and the investigator can make a relatively simple circuit board to operate them.

THURTELL: In Canada, we purchase the chips for about $80.00.
We also make our own units to operate the chips.

COYNE: I don't believe coated sensors are available any more.
We were not able to obtain one recently. We have a humidity circuit that we have designed to operate this sensor that is available from our lab, the Lawrence Livermore Laboratory. The important consideration is to get electronics as close as possible to
the sensor because capacitance change in the sensor (0.2 pico-
farad/1% RH). Otherwise capacitance changes in the leads can have
significant effects on the indicated relative humidity.

PAIGE: Sensors purchased directly from Finland cost $50.00
and when mounted on a circuit board they cost $100.00. Some of
the reliability problems in these units are related to the fact
that if they are adjusted with a metal screwdriver, the probe is
shorted out. The sensors are made of glass and hence are fragile.
However, we found they are electrically and mechanically very
stable under continuous use.

SEARLS: As manufacturers, we have been greatly concerned with
durability in humidity controlling systems. The dew cell system
and the wet bulb-dry bulb systems have been the most durable. We
have been using a new wet bulb dry bulb system that is electronic
with a signal conditioner built in and made up with a microprocessor. These have proved to be very durable over the 1-1/2
years that we have had them in use.

CURRY: Hoffman, why didn't you refer to moisture measurement
in terms of $g\ g^{-1}$ in your presentation?

HOFFMAN: It is more difficult to measure moisture as a mass
measurement compared to other commercial techniques. There is
no reason why the investigator cannot convert from any other
moisture measurement as long as he has temperature and pressure
measurements at the time he makes the moisture measurement.

THURTELL: Any absorption measurement as infrared or Lyman-
alpha technique does measure the density of water vapor in the
air.

CARBON DIOXIDE

J. E. Pallas, Jr.

Southern Piedmont Center, USDA-SEA
Watkinsville, Georgia

INTRODUCTION

An investigator planning to control CO_2 concentration in a
controlled environment might first ask what the normal CO_2 concentration is in the terrestrial environment. At present it is about
330 ± 6 µl CO_2 l^{-1} air but the actual concentration depends somewhat on when and where it is measured.[1] For investigators working
in controlled environments it is of some physiological significance to know that the CO_2 concentration of the atmosphere has
been steadily increasing for the last few hundred years and will
probably continue to do so. The CO_2 content of the atmosphere
has risen from near 275 µl l^{-1} before the industrial revolution
(Bray, 1959; Bolin and Keeling, 1963; Stuiver, 1978) to near 330
µl l^{-1}, based on observations of Keeling et al. (1976) at Mauna
Loa Observatory in Hawaii (Fig. 1). The Mauna Loa measurements
indicate that the average CO_2 content of the atmosphere has risen
by more than 5% since 1958. Similar trends have been shown for
Scandinavia (Bolin and Bischop, 1970) and Antartica (Keeling et
al., 1976). The increase, which has been exponential since the
preindustrial revolution, is considered to be largely the result

[1]Concentrations can be converted to millimoles m^{-3} by multi-
plying µl CO_2 l^{-1} air by 0.0416.

of burning of fossil fuels, production of cement (Keeling, 1973),
defoliation of forests, and burning of wood and peat (Adams, Mantovoni, and Lundell, 1977; Bolin, 1977). Thus it is conceivable
that the global mean atmospheric CO_2 concentration may exceed 600
ppm sometime in the next century (Bacastow and Keeling, 1973;
Baes et al., 1977). These trends emphasize that the appropriate
CO_2 level in a controlled environment facility is not a fixed
standard and theoretically might be based on the decade to which
the experimenter wishes to relate his studies. A seasonal fluctuation of approximately 6 µl CO_2 l^{-1} air can be noted in the
Mauna Loa data (Fig. 1). This reflects the yearly trend of high
photosynthesis during the spring and summer and its decrease in
fall and winter in the Northern Hemisphere.

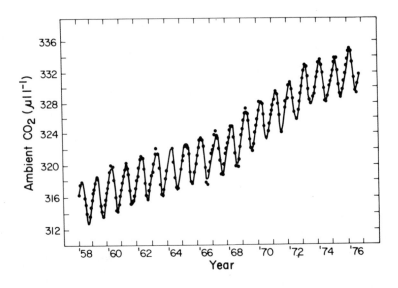

FIGURE 1. Variation in atmospheric CO_2 concentration as measured at Mauna Loa Observatory, Hawaii. (From Keeling et. al.,
1976).

From measurements we have taken from a mobile unit in the
Athens, Georgia area, (Fig. 2), we have concluded that the ambient

CO_2 concentration in an urban area may at times exceed 400 μl l^{-1} due to fuel consumption.

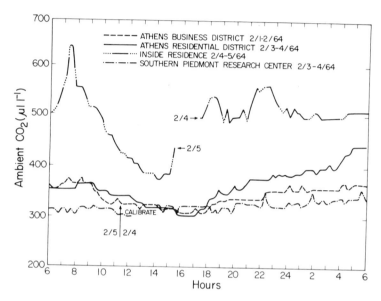

FIGURE 2. Ambient CO_2 concentrations as measured at several locations in the Athens, Georgia area.

CO_2 VARIATIONS WITHIN PLANT GROWTH CHAMBERS

During photosynthesis, CO_2 molecules are continuously extrac-
ted from the air surrounding the plant. In the dark the situation
is reversed, with respired CO_2 released by the plant. In an almost air-tight controlled environmental chamber, if the CO_2 depleted by photosynthesis is not replenished, the CO_2 concentration
during the light hours may decrease to a level well below that of
ambient air outside the chamber, and during the dark it may well
increase above that of the ambient air. On the other hand, in an
open system with supplementary air flow through the chamber and

leakage from the laboratory area, a chamber CO_2 concentration somewhat higher than that considered normal for the terrestrial environment may occur even during the light hours because of elevated levels of CO_2 in areas where people are working and breathing.

Problems with CO_2 fluctuations in controlled environment facilities were first brought to the attention of physiologists in 1963 (Pallas, 1963). At that time it was recognized that "the control and monitoring of CO_2 concentrations have been found to be essential in an environment where temperature, light and humidity are held constant." This warning was based on a number of observations. For example, measurements of CO_2 content in a high-light controlled environment room at Watkinsville, Georgia (Williams et al., 1961) during a study of corn plants (Zea mays L.) revealed that CO_2 content in the controlled environment room frequently ranged from 350 to more than 400 μl CO_2 l^{-1} air (Fig. 3) when the ambient CO_2 concentration out of doors was 315 μl l^{-1} (Pallas et al., 1965). Changes in the rate of photosynthesis of corn plants were also positively associated with fluctuations in the CO_2 concentration in the growth room (Fig. 3). Other studies indicated that the higher than ambient CO_2 concentration in the growth room was inducing stomatal closure (Pallas and Bertrand, 1966). The additional CO_2 was traced to several sources: investigators in the growth room, people in nearby laboratories and offices, gas fired furnaces and water heaters, and vehicles in an adjacent parking lot. Unfortunately there is a tendency for investigators to place their controlled environment facilities in the most convenient locations when facility location should be a matter of practicality concerning CO_2 control. Many chambers are in poorly ventilated, high CO_2 areas, an unfortunate choice of location. Location of chambers in areas that have considerable human activity and minimal fresh air intake will lead to large fluctuations in chamber CO_2, especially when chamber doors are opened frequently. In attempts to significantly lower the CO_2

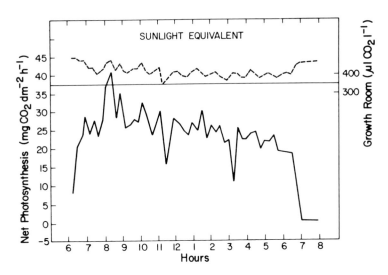

FIGURE 3. *Net photosynthesis of corn plants (solid line) and CO_2 concentration (broken line) in the high light controlled environment rooms at Watkinsville, Georgia.*

concentration in our growth room, investigators were required to wear masks that were connected by flexible tubing to the out of doors. This was not as successful as we hoped it would be in lowering and smoothing out chamber CO_2 concentration. In studies with cotton (*Gossypium hirsutum* L.) the use of face masks made it possible to maintain CO_2 concentraions on the average below 350 µl CO_2 l^{-1} air; peaks in CO_2 were still associated with intense activity within the room (Fig. 4) (Pallas, Michel, and Harris, 1967). Fluctuations in chamber CO_2 were again reflected in differences in rates of photosynthesis. Monitoring of CO_2 in small tight growth chambers indicated that during daylight hours the CO_2 concentration stayed below that of ambient air except when the doors were open or investigators were inside the chamber. French, Hiesey, and Milner (1959) had earlier warned that CO_2 depletion could be a problem in growth chambers. We found that the rate of depletion and lowest CO_2 concentrations experienced depended greatly on plant size and density as well as plant

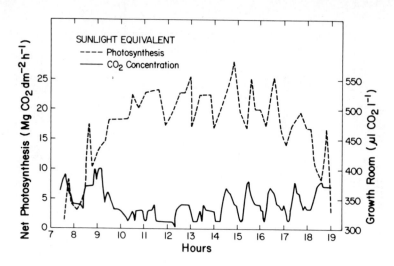

FIGURE 4. Net photosynthesis of cotton plants and CO_2 concentrations in the high light controlled environment room at Watkinsville, Georgia.

species or mixture of species in the chamber. In our almost airtight chambers with a large C_3 plant population, CO_2 compensation point concentrations, approximately 50 µl CO_2 l^{-1} air, were obtainable in a matter of hours unless CO_2 was injected into the chamber. Since a liter of air containing 300 µl CO_2 at standard temperature and pressure holds only 0.6 mg CO_2, photosynthetic depletion of CO_2 can occur relatively rapidly. The CO_2 response curves for the C_3 plant peanut (*Arachis hypogaea* L.) at two light levels (Fig. 5) demonstrate the dependency of photosynthesis on CO_2 concentration and thus the importance of maintaining CO_2 level. Because our chambers were air tight, they were eventually used as semi-closed systems for measuring photosynthesis (Pallas, 1973). Low CO_2 concentrations of air can also have a very significant effect on plant water balance by increasing stomatal opening (Pallas, 1965). In general, experience has shown that change in CO_2 concentrations is influenced largely by the number of people in and around growth chambers, time of day, daylength, and time of year.

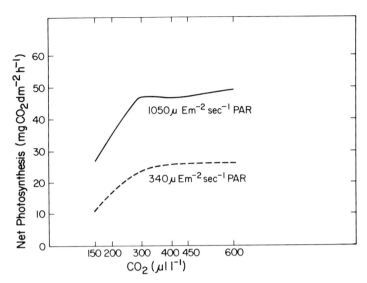

FIGURE 5. *Net photosynthesis of peanut plants at two light levels as influenced by CO_2 concentration.*

Routine control of CO_2 in controlled environmental facilities has rarely been mentioned in the literature. Michel (1977) measured CO_2 levels outside a growth chamber to be well above 400 $\mu l \ CO_2 \ l^{-1}$ air during nonexperimental periods. To maintain a consistent CO_2 level, he automatically supplemented the chamber air with CO_2 to maintain 500 $\mu l \ l^{-1}$ during nonexperimental periods. He stated that he controlled CO_2 at 300 $\mu l \ CO_2 \ l^{-1}$ during most of the experimental periods. Recommendations for CO_2 control have been made more recently by Tibbitts and Krizek (1978).

Patterson and Hite (1975) reported on CO_2 measurements in phytotrons where introduction of outside air (1% of a chamber volume per minute) failed to maintain the CO_2 concentration in plant chambers. Cotton plants lowered the concentration to 150 $\mu l \ CO_2 \ l^{-1}$ and corn to 50 $\mu l \ CO_2 \ l^{-1}$, although outside air contained 350 $\mu l \ CO_2 \ l^{-1}$. I. J. Warrington at the DSIR Climate Laboratory, Palmerston North, New Zealand,[2] indicated that large corn

[2] *personal communication*

plants depleted CO_2 to 100 µl CO_2 1^{-1} within 60 to 90 minutes in
growth rooms with four air exchanges per hour or less. By comparison, CO_2 concentration exceeded 500 to 600 µl CO_2 1^{-1} air
when one investigator stayed in the growth room for several minutes. The decay rate was slow- 1 to 2 hours with small plant
loads. From studies on single plants Talbot (1970) estimated
that a room filled with plants under optimum conditions would
consume CO_2 at a maximum rate of 1 gm per minute. Since each of
his rooms had a total volume of 38 kl, the rate of change in concentration was estimated at 13 µl CO_2 1^{-1} air per minute without
supplementary air. No mention was made of night time CO_2 concentrations.

Regardless of the amount of fresh air added to small plant
growth chambers there normally will be some CO_2 depletion during
the light period and an increase during the dark period that automatically introduces a CO_2 cycle in combination with the light
dark cycle. Tibbitts and Krizek (1978) measured the magnitude of
the fluctuation and calculated the equilibrium CO_2 levels in a
reach-in chamber as affected by air exchange (Table 1).

CO_2 AND PLANT RESPONSE

Strain (1978) listed 32 plant responses to CO_2 enrichment
(Table 2). Allen (1979) reviewed the potential for CO_2 enrichment and the reader is referred to his paper for further information. A topic not previously covered, however, is the need for
CO_2 control in growing plants for biochemical studies. We have
studied the level of a key enzyme in photosynthesis and photorespiration of peanut leaves; we questioned whether the ambient
CO_2 level during growth of test plants could affect enzymatic
activity. Peanut plants were grown to 4 weeks of age in separate
growth chambers under three levels of CO_2: 150, 300, or 450 µl
CO_2 1^{-1} air as previously described (Pallas, Samish, and Willmer,

TABLE 1. *Carbon Dioxide Levels in a Reach-in Chamber as Affected by Fresh Air Exchange. (From Tibbitts and Krizek, 1978).*

Additions of fresh air of 320 μl CO_2 l^{-1} air	Equilibrium CO_2 level	
chamber volumes per min	μl l^{-1}	
	light[a]	dark[b]
1.0	313	324
.9	312	324
.8	311	325
.7	310	325
.6	308	326
.5	306	327
.4	302	329
.3	296	332
.2	284	338
.1	248	356

[a]*Net photosynthetic rate of 13.7 mg CO_2 hr^{-1} dm^{-2} plant-bed surface area butterhead lettuce at 20° C, 80% RH, and 32.5 μE cm^{-2} s^{-1} plant bed surface of 39 dm^2.*

[b]*Net respiration rate of 6.8 mg CO_2 hr^{-1} dm^{-2} plant bed surface area butterhead lettuce at 20° C and 80%, RH.*

1974). Assays were made of D-ribulose-1, 5-bisphosphate carboxylase-oxygenase activity as well as protein and chlorophyll contents (Lorimer, Badger, and Andrews, 1977; Bradford, 1976; Arnon, 1949). The low CO_2 treatments affected fresh weight. (Table 3). Fresh weight increased with increasing CO_2 level as did protein, but only slightly. However, chlorophyll concentration was considerably lower at 450 μl CO_2 l^{-1} air than at 300 μl l^{-1}. D-ribulose-1, 5-bisphosphate carboxylase activity and the RuBPC/oxygenase

ratio decreased progressively with increase in CO_2 concentration.
Thus, it is evident that the CO_2 concentration in plant growth
chambers may not only influence dry weight increment and photosynthetic rate but also protein and chlorophyll contents, as well
as enzyme activity. These observations further the argument for
monitoring and maintaining of CO_2 levels in controlled environment
facilities.

TABLE 2. *Plant Responses to CO_2 Enrichment. (From Strain, 1978).*

1. *Photosynthesis increase*
2. *Photorespiration change*
3. *Dark respiration change*
4. *Photosynthate compostiion change*
5. *Photosynthate allocation change*
6. *Growth rate change (dry weight and elongation)*
7. *Transpiration decrease*
8. *Stomatal conductance decrease*
9. *Leaf temperature increase*
10. *Tolerance to atmospheric pollutant increase*
11. *Leaf area increase*
12. *Leaf dry weight/leaf area increase*
13. *Leaf senecence rate change*
14. *Stem diameter decrease*
15. *Node number increase*
16. *Lateral branches increase in number*
17. *Root/shoot increase*
18. *Cytological changes*
19. *Flowers produced earlier*
20. *Flower size decrease*
21. *Number of flowers change*
22. *Fruit size increase*
23. *Fruit number increase*
24. *Accelerated maturity of crop*

TABLE 2. (continued)

25. Seeds/plant increase

26. Germination of seeds and spores induced

27. Dark CO_2 fixation increase

28. Effects on CAM plants

29. Growth rates of heterotrophic organisms increase

30. Symbiotic and non-symbiotic nitrogen increase

31. Inter- and intraspecific differences in magnitude of various responses

32. Interspecific differences in nature of various responses

MEASUREMENT OF CO_2 IN CONTROLLED ENVIRONMENTS

Bailey et al. (1970) reviewed methods of monitoring CO_2 in confined structures such as greenhouses and growth chambers. The sensors of the monitoring systems depend on absorption of infrared radiation and on electrochemistry, photochemistry, interferometry, gas chromatography, and liquid scintillation spectrometry. Infrared radiation absorption is used almost exclusively in controlled environments with a small number of laboratories using electrochemistry. These two systems will be discussed briefly while recognizing that other systems may also be usable but have limitations for continuous monitoring in controlled environments. Infrared gas analyzers are more expensive and complicated than instruments based on electrochemistry such as the conductimetric CO_2 analyzer. However, an infrared gas analyzer (IRGA) has the potential for much greater precision in monitoring CO_2. Its main advantage is that it has capacity for continuous and direct measurement of CO_2 concentration, it can be used directly with recorders and control devices, and its output can interface directly with a computer. Companies that manufacture conductimetric and IRGA units are listed in Table 4.

TABLE 3. Effect of Ambient CO_2 Level During Growth on Certain Constituents of Peanuts.

Chamber CO_2 Concentration $\mu l\ l^{-1}$	Fresh Weight g	RuBPC[a]	RuBPO[b] $\mu mole\ mg\ prot^{-1}\ min^{-1}$	$\dfrac{RuBPC}{RuBPO}$	Protein Dry Weight $mg\ g^{-1}$	Chlorophyll a Dry Weight $mg\ g^{-1}$	Chlorophyll b Dry Weight $mg\ g^{-1}$	a/b
150	3.79	.4598	.039	11.79	47	10.6	8.4	1.3
300	16.93	.4199	.036	11.66	48	11.6	9.2	1.3
450	18.84	.3921	.036	10.89	51	6.4	5.0	1.3

[a] D-ribulose-1, 5-bisphosphate carboxylase

[b] D-ribulose-1, 5-bisphosphate oxygenase

TABLE 4.   *Manufacturers of Conductrimetric and Infrared Gas Analyzers.*

Conductimetric

Hampden Test Equipment Limited
Rothersthorpe Ave.
Northhampton, NN4 9JH U.K.

IRGA

Anarad, Inc.
P. O. Box 3160
Santa Barbara, California 93105 USA

Beckman Instruments, Inc.
2500 Harbor Blvd.
Fullerton, California 92634 USA

Esterline Angus Instrument Corporation
Box 24000
Indianapolis, Indiana 46224 USA

Fuji Electric Corporation of America
Room 927, 30 East 42nd St.
New York, New York 10017 USA

G. P. Instrumentation NE I Electronics Ltd.
Whitley Road, Longbenton
Newcastle upon Tyne, NE12 9SR U.K.

Hartman & Braun A6
Grafstrasse 97
Postfach 900507, 6 Frankfurt/Main,90 Germany (BRD)

V. E. B. Junkalor Dessau
Altener Str. 43, 45
Dessau, Germany (DDR)

H. Maihak AG
Semperstrasse 38,
Postfach 601709
2000 Hamburg 60, Germany (BRD)

M.S.A. Research Corporation
600 Penn. Center Blvd.
Pittsburg, Pennsylvania 15235 USA

TABLE 4. (continued)

Sereg Schlumberger
100 Rue De Paris - BP65
91302 Massey Cedex, France

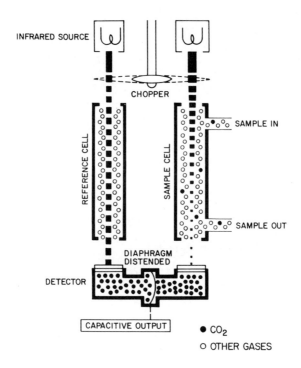

FIGURE 6. Functional diagram of a typical infrared gas analyzer showing movement of sample gas and infrared radiation path.

IRGA SENSING SYSTEMS

The IRGA operates on the principle that the CO_2 molecule absorbs energy in the infrared region of the electromagnetic spectrum. The essentials of an IRGA (Fig. 6) consist of (1) a suitable source of infrared radiation usually provided by two balanced tungsten filaments, (2) two columnar cells which are fitted with

appropriate windows, one cell of which is filled with or allows
for flow-through of standard gas such as nitrogen and the other
cell for sample gas. The internal surfaces of the cells normally
are coated with gold foil so as to efficiently reflect infrared
energy, (3) a mechanical chopper to modulate infrared radiation
from the source, (4) a microphone or a "luft chamber". Figure
6 is representative of a system used in conjunction with a chopper
design that allows concurrent passage of sample and reference
radiation. The chambers of the luft detector are charged with
the gas to be detected or suitable look-alike gas. The gas in
the chambers absorbs the radiation remaining after passage
through the sample and reference cells. Difference in energy due
to sample absorption is sensed by a flexible metal diaphragm
which yields a capacitive output and is interpreted as a quantity
of gas in the sample cell. The absorption bands of CO_2 and H_2O
overlap in the region near 2.7 μm. Hence, measurement of CO_2 in
trace concentrations by an IRGA can be affected by water vapor if
all absorption bands are used unselectively. At ambient temperature of 25°C and a barometric pressure of 760 mm Hg we have measured a difference of 10 μl CO_2 l^{-1} air by a Model 215A-S Beckman
IRGA[*] between dry and saturated air at a concentration of 300 μl
CO_2 l^{-1} air. Normal humidity fluctuations are far less and would
approximate the routine error at ambient CO_2 levels under constant
conditions of $\pm$ 3μl CO_2 l^{-1} that we have noted when the IRGA was
operating properly. On the other hand our CO_2 injection system
only controls routinely to $\pm$ 15μl CO_2 l^{-1} air with a moderate
noise level. All changes which affect CO_2 density in the IRGA
cells affect its response; thus barometric pressure and temperature
may have considerable effect. For more complete details concerning
IRGA stability, precision, and calibration the reader is referred
to Sestak, Jarvis, and Catsky (1971).

*Trade names and company names are given for benefit of reader
and do not imply preferential treatment by the U.S. Department of
Agriculture.

ELECTROCHEMICAL SENSING SYSTEMS

The conductimetric CO_2 analyzer operates on the principle
that the electrical conductivity of water increases with increasing dissolved CO_2 concentration. Electrical conductivity is proportional to CO_2 concentration. Simply, an air sample is pumped
through a bubble column containing deionized water. During the
bubbling process CO_2 from the air sample dissolves in the water,
thereby increasing its conductivity. The water from the bubble
column flows to a cell which measures the electrical conductivity.
In the most common closed loop circuit the water passes through
a deionizer and back to the bubble chamber, to give a continuous
measuring process. Major drawbacks to this type of analyzer are
its limited accuracy (± 15 μl l^{-1} at 300 μl CO_2 l^{-1} air) and temperature sensitivity (30 μl CO_2 for each $^\circ C$ change). However, a
modification of this type of instrument with a circuit employing
a thermistor has recently been shown to reduce temperature sensitivity to approximately 1 μl CO_2 for each $^\circ C$ (Kimbell and Mitchell, 1979).

CONTROL OF CO_2 IN CONTROLLED ENVIRONMENTS

We have been concerned primarily with monitoring and controlling CO_2 during the light hours in individually controlled environmental chambers. An IRGA has been routinely used in our laboratory for over 15 years. The chamber air is sampled at the center of the outlet to the chamber air exchange system and pumped
through the IRGA by a small bellows or diaphram pump. The output
from the analyzers is directed to meters with contact switches
that operate solenoid valves to provide CO_2 flow from cylinders
of compressed gas. The needle valve with a flow meter provides
the final control for bleed-in to the chamber. CO_2 is injected
in the inlet port from the air exchange system. Essentially the
same basis for a system is described by Tibbitts and Krizek (1978).

Since our chambers have an average air velocity of 25 m min^{-1}, good mixing is assured. When it is desirable to maintain below ambient concentrations of CO_2 during the light period, we have grown populations of seedling corn, sorghum or millet in the chambers to absorb CO_2 which can effectively keep the CO_2 concentration low or at some set point of the injection system. Nighttime CO_2 has been controlled by connecting two controlled environmental chambers together with stove pipe and running them on opposing light schedules, with the chamber with lights acting as a scrubber for the other chamber over its dark period. Our attempts to develop chemical scrubbing systems for controlled plant growth facilities have not been successful. The use of the caustic substances provided only limited capacity for scrubbing and caused a cleanup problem. It is reported that commercial scrubber units are being used in Japan (T. W. Tibbitts[1]).

Our most recent CO_2 controller is comprised of comparator circuitry designed by the University of Georgia electronics shop. We have used such units without problems since 1973. Scientific Systems Corporation, Baton Rouge, Louisiana markets a CO_2 recorder-controller system utilizing an IRGA. Supplementary CO_2 is automatically pulsed to the controlled environment to maintain a preset carbon dioxide level. The manufacturer's specifications indicate that the typical control span is $\pm$ 1% of the analyzer scale.

Hampden Test Equipment Limited, U. K. (Table 4) markets a Gas-O-Mat Indicator/Controller system based on the conductrimetric principle. A gas switch is also available which enables one analyzer to be used to monitor CO_2 levels in up to 10 chambers. Accuracy is reported as $\pm$ 5% for full scale deflection of 1000 μl CO_2 l^{-1}.

[1]*Personal communication*

MULTIPLE CHAMBER CONTROL

Patterson and Hite (1975) described a system used to continuously monitor and supplement CO_2 in several growth chambers and greenhouses of the Duke University phytotron. In the original system, air samples were drawn continuously from the chambers through heated tubing (to prevent condensation) to a solenoid valve at the control panel. A stepping switch activated every 2 minutes selected the sample to be analyzed through a Beckman Model 864 IRGA with readout on a recorder. A two-way normally closed solenoid valve at each chamber controlled a needle valve with flow meter for the input of CO_2 from a compressed gas cylinder. The solenoid was activated for 2-minute injections by a cam activated switch from a recorder open drive mechanism. This CO_2 injection system generally maintained the CO_2 concentration within $\pm$ 25 µl l^{-1} of the set point throughout the day. The system was originally tested using a sampling interval of 14 min. However, it was suggested that use of a shorter sampling interval could decrease fluctuations of CO_2 during each cycle. This was because during each pulse enough CO_2 had to be injected to last until the next pulse. This CO_2 monitor and controller system has since been updated at the Duke University phytotron. Samples from chambers are now selectively fed through an IRGA and a small computer runs a comparison between the sample and set point for each chamber. If the reading is lower or higher than the set point, the injection rate is raised or lowered until reset during the next sampling. With all chambers on line, a complete cycle takes almost 20 min. Fluctuations in control of $\pm$ 15 µl CO_2 l^{-1} air have been recorded with the set up. The reading time of any one chamber can be shortened from 1 minute to 30 seconds or chambers can also be bypassed by programming.

Talbot (1970) described the CO_2 control system in use at the phytotron at Palmerston North, New Zealand. This is a rather sophisticated system that uses three Hartman and Braun URAS 2 infrared analyzers to measure CO_2 concentration in 24 controlled rooms and a fourth to monitor for malfunctions. Each control IRGA is coupled to eight rooms, sampling sequentially one room per minute. At the end of each sample period, the output of the IRGA is compared with the set point for a room and the error is digitized and stored in a 4 bit reversible counter associated with the room. Every 30 seconds the counter is interrogated and, if the count is not zero, CO_2 is added to the room in a pulse. Carbon dioxide is injected during the period between sampling times, and the length of each pulse may be varied to suit plant loads. Each room is also sampled for 10 minutes every 4 hours by the monitor analyzer. If the CO_2 content of a room reaches levels above or below the variance limits of the setting, an alarm is generated through the alarm discriminator. If any of the control analyzers record a sudden large excess of CO_2 they activate an alarm system.

Bailey *et al.* (1970) described a CO_2 control system used in the Plant Stress Laboratory at Beltsville, Maryland, involving an IRGA that was engineered to control up to twelve growth chambers.

Every biotron, phytotron controlled environment facility has its own potential for control. In some instances fresh air intake may suffice. At the University of Wisconsin Biotron, CO_2 is controlled by introducing large volumes of make-up air.

Although control of CO_2 in the dark in single or multiple chamber arrangements could be of some advantage, no simple, adaptable working system is available at the present time. This is a problem area that could use some engineering ingenuity.

REFERENCES

Adams, J. A. S., Mantovoni, M. S. M. and Lundell, L. L. (1977). Wood versus fossil fuel as a source of excess carbon dioxide in the atmosphere: A Preliminary Report. *Science 196*,54-56.

Allen, L. H., Jr. (1979). Potentials for carbon dioxide enrich-
    ment. *In* "Modifying the Aerial Environment of Plants" (B.
    J. Barfield and J. F. Gerber, eds.), pp. 500-519. Monograph
    Amer. Soc. Agric. Eng., St. Joseph, Michigan.

Arnon, D. I. (1949). Copper enzymes in isolated chloroplasts.
    Polyphenoloxidase in *Beta vulgaris*. *Plant Physiol. 24*,1-15.

Bacastow, R., and Keeling, C. D., (1973). Atmospheric carbon di-
    oxide and radiocarbon in the natural carbon cycle: II.
    Changes from AD 1700 to 2070 as deduced from a geochemical
    model. *In* "Carbon and Biosphere" (G. Woodwell and E. Peron,
    eds.), pp. 86-134. U. S. Atomic Energy Commission, Washington
    D. C.

Baes, C. F., Jr., Goeller, H. E., Olson, J. S., and Rotty, R. M.,
    (1977). Carbon dioxide and climate. The uncontrolled exper-
    iment. *Amer. Scientist 65*,310-320.

Bailey, W. A., Klueter, H. H., Krizek, D. T., and Stuart, N. W.,
    (1970). CO_2 systems for growing plants. *Trans. ASAE 13*,263-
    268.

Bolin, B., and Bischop, W., (1970). Variations in the carbon di-
    oxide content of the atmosphere in the Northern Hemisphere.
    Tellus 22,431-442.

Bolin, B., (1977). Changes of land biota and their importance
    for the carbon cycle. *Science 196*,613-615.

Bolin, B., and Keeling, C. D., (1963). Large scale atmospheric
    mining as deduced from the seasonal and meridional varia-
    tions of carbon dioxide. *J. Geophys. Res. 68*,3899-3920.

Bradford, M. M. (1976). A rapid and sensitive method for the
    quantitation of microgram quantities of protein utilizing
    the principle of protein-dye binding. *Anal. Biochem. 72*,248-
    254.

Bray, J. R. (1959). An analysis of the possible recent change
    in atmospheric carbon dioxide concentration. *Tellus 11*,220-
    230.

French, C. S., Hiesey, W. H., and Milner, H. W., (1959). Carbon
    dioxide control for plant growth chambers. pp.352. Carnegie
    Inst. Washington Yearbook (1958).

Keeling, C. D. (1973). Industrial production of carbon dioxide
    from fossil fuels and limestone. *Tellus 25*,174-198.

Keeling, C. D., Bacastow, R. B., Bainbridge, A. E., Ekdahl, C. A.
    Jr., Guenther, P. R., Waterman, T. S., and Chin, J. F. S.
    (1976). Atmospheric carbon dioxide variations at Mauna Loa
    Observatory, Hawaii. *Tellus 28*,538-551.

Keeling, C. D., Adams, J. A. Jr., Ekdahl, C. A. Jr., and Guenther,
    P. R., (1976). Atmospheric carbon dioxide variations at the
    South Pole. *Tellus 28*,552-564.

Kimbell, B. A., and Mitchell, S. T., (1979). Low-cost carbon di-
    oxide analyzer for greenhouses. *Hortscience, 14*,180-182.

Lorimer, G., Badger, N., and Andrews, T., (1977). D-ribulose-1,
    5-bisphosphate carboxylase-oxygenase. *Anal. Biochem. 78*:66-
    75.

Michel, B. E. (1977). A model relating past permeability to flux
    and potentials. *Plant Physiol 60*,259-264.

Pallas, J. E., Jr. (1963). CO_2 and controlled environmental re-
    search. *Plant Physiol. 38*,xxxv.

Pallas, J. E. Jr., (1965). Transpiration and stomatal opening
    with changes in carbon dioxide content of the air. *Science
    147*,171-173.

Pallas, J. E., Jr., Bertrand, A. R., Harris, D. G., Elkins, C. B.
    Jr., and Parks, C. L., (1965). Research in plant transpira-
    tion: 1962. Prod. Res. Rpt. 87, Agric. Res. Ser. USDA.

Pallas, J. E., and Bertrand, A. R., (1966). Research in plant
    transpiration: 1963. Agr. Prod. Res. Rpt. 89. Agric. Res.
    Ser. USDA.

Pallas, J. E., Michel, B. E., and Harris, D. G., (1967). Photo-
    synthesis, transpiration, leaf temperature and stomatal ac-
    tivity of cotton plants under varying water potentials.
    Plant Physiol. 42,76-88.

Pallas, J. E., Jr. (1973). Diurnal changes in transpiration and daily photosynthetic rate of several crop plants. *Crop. Sci.* *13,82-84.*

Pallas, J. E., Jr., Samish, Y. B., Willmer, C. M., (1974). Endogenous rhythmic activity of photosynthesis, transpiration, dark respiration, and carbon dioxide compensation point of peanut leaves. *Plant Physiol.* *53,*907-911.

Patterson, D. T., and Hite, J. L., (1975). A CO_2 monitoring and control system for plant growth chambers. *Ohio J. Sci. 75:* 190-193.

Sestak, Z., Jarvis, P. G., and Catsky, J., eds. (1971). "Plant Photosynthetic Production - Manual of Methods." N. Junk, The Hague.

Strain, B. R., ed. (1978). Report of the workshop on anticipated plant responses to global carbon dioxide enrichment. Duke Environmental Center, Duke University, Durham, North Carolina.

Stuiver, N. (1978). Atmospheric carbon dioxide and carbon reservoir changes. *Science 199,*253-258.

Talbot, J. S. (1970). A multiplexed digital control system. *Radio, Electronics and Communications 25,*21-23.

Tibbitts, T. W., and Krizek, D. T., (1978). Carbon Dioxide. *In* "A Growth Chamber Manual" (R. W. Langhans, ed.), pp. 80-100. Cornell University Press, Ithaca, N.Y.

Williams, G. G., Pallas, J. E., Jr., Harris, D. G., Elkins, C. B., Jr. (1961). Research in plant transpiration. U. S. Army Electronics Proving Ground. Fort Huachuca, Arizona.

CARBON DIOXIDE: CRITIQUE I

Henry Hellmers
Lawrence J. Giles

Phytotron
Duke University
Durham, North Carolina

Pallas' opening paper on CO_2 described the problems encountered in growing plants in controlled environments. He also went into detail on methods of measuring the CO_2 concentration. Another problem he mentioned, and the one that will be expanded here, is measurement and control of CO_2 in multiple chambers.

The need for CO_2 regulation in controlled environment plant growth chambers depends on the relation of the size of the chamber to that of the size, number, and type of plants being grown in the chamber. Studies in both units of the Southeastern Plant Environment Laboratories, Duke University and North Carolina State University, have demonstrated that C_3 and C_4 type plants can lower the CO_2 level in chambers to 150 ppm and 50 ppm, respectively. It has been shown that the decline in CO_2 level occurred, even with make-up air, at a rate of 1% of chamber volume per minute (Patterson and Hite, 1975). CO_2 concentration in a fully loaded chamber containing tobacco plants, C_3, decreased by 200 ppm in 30-40 minutes (Downs and Hellmers, 1975). Furthermore, the CO_2 content of a chamber filled with plants increases rapidly when an investigator enters a chamber. Also, during the night the CO_2 concentration increases as a result of plant respiration.

229

To control the level of CO_2, especially if it is desired to only maintain a level equivalent to that of outside air, the first thought is to increase the air exchange rate of the chamber. However, to prevent a rapid decline in CO_2 level in chambers filled with large rapidly photosynthesizing plants would require an air input rate that replaces 75% of the air every minute (Downs and Hellmers, 1975). Such an exchange rate makes it virtually impossible to maintain temperature control if the desired temperature is different from that of air which is being circulated into the chamber.

MAINTAINING FIXED CO_2 CONCENTRATION

The problem posed by a chamber that contains photosynthesizing plants that cause the CO_2 concentration to decrease below a desired level can be best solved by adding compressed CO_2. The problem is to determine how much CO_2 is required per unit time and then to add just that amount. While the methods to solve the problem are well known, the equipment is expensive, particularly for the CO_2 analyzer. The cost is lowered per chamber if a system is developed to control several chambers with one analyzer but this introduces the additional problem of precise control between sampling times.

The necessary equipment includes an infrared gas analyzer (IRGA) or other CO_2 measuring system along with assorted flow meters, pumps, and plumbing to determine the CO_2 concentration in the chamber. Supplementary CO_2 is provided to chambers through release of CO_2 from compressed gas cylinders into the chambers. A relatively straightforward system that has worked especially well in the Duke Phytotron for up to five chambers, was described by Patterson and Hite (1975). In their system a cam switch on the recorder activated a solenoid valve when the measured CO_2 level dropped below a preset concentration. The solenoid valve controlled the flow from a tank of compressed CO_2

and could be held open until the CO_2 level in the chamber reached
the predetermined level.

The system has now been modified to include additional chambers, to control different chambers at different levels of CO_2
concentration, and to reduce the degree of fluctuation in the
CO_2 concentration between sampling times. The same sampling
system and the injection equipment described by Patterson and
Hite (1975) was used.

A limitation encountered in maintaining a relatively constant
level of CO_2 in multiple units was the sampling time. To purge
the tubing and the analyzer and to get an accurate reading,
especially when large differences in CO_2 levels were used in
different chambers, required a minimum of 30 seconds. This
meant that the 20 chambers and six temperature controlled greenhouses could be sampled only once every 14 minutes. One minute
was used to calibrate the IRGA system using a standard gas. To
overcome the problem of decline in CO_2 concentration in the units
between sampling times, and to obtain different CO_2 concentrations in different units, a microcomputer was put into the system
in place of the recorder that activated the input valves.

The microcomputer performs the following functions:

1. Compares the CO_2 level measured by the IRGA for each
   controlled environment unit with a preprogrammed level
   of CO_2 for that unit.

2. Calculates an injection rate of CO_2 in terms of the
   number of seconds that injection valve has to be open
   each minute to maintain the required levels.

3. Controls the opening of the injection valve during the
   time between samplings. Injection is with 0.5 second
   pulses and can be controlled for 0 to 9 pulses during
   each minute between sampling times.

4. Adjusts the calibration of the IRGA system to corres-
   pond to the calibration gas.

A record of the injection rate and the CO_2 level for each
unit is obtained on punch tape at present. Eventually it will
be recorded on disks.

Currently, the system can maintain each of the 26 units at
any desired level of CO_2 concentration between 350 and 1000 ppm.
For chambers requiring higher levels of CO_2 the injection valve
opening can be set manually to allow a greater volume of gas to
enter the chamber with each injection. Most of the time while
the lights are on, the fluctuation in CO_2 is held within ±10 ppm
when the desired concentration is 350 ppm (Fig. 1). However, at
higher concentrations the fluctuation increases. One problem
currently being corrected is an overshoot in the declining CO_2
concentration when the lamps are turned on and the CO_2 level is
high. A second problem is an upward overshoot when an investigator enters the chamber. The first problem is caused by the
fact that the CO_2 system is a reaction system and is expected to
be corrected by programming into the microcomputer, an anticipa-
tion factor that is activated by turning on the lights. The
second problem cannot be completely solved unless investigators
wear CO_2 absorbing masks because people emit so much CO_2.
However, a partial solution can be obtained by interlocking the
door and the injection system so that when the door is opened no
additional CO_2 will be injected until the next time the air is
analyzed.

The maintaining of chambers at CO_2 levels below ambient
involves an entirely different set of problems. Due to the
respiration of the plants plus leakage into the chambers, a
system would have to be developed to continuously remove CO_2
from the atmosphere. This system would have to handle a large
volume of air because the carbon dioxide content is relatively
small compared to the other gas components. There are two methods
of removing the CO_2 that are feasible but both would be expensive.
One method would involve absorption of CO_2 on columns or in
alkaline solutions. This would require an investment in equip-

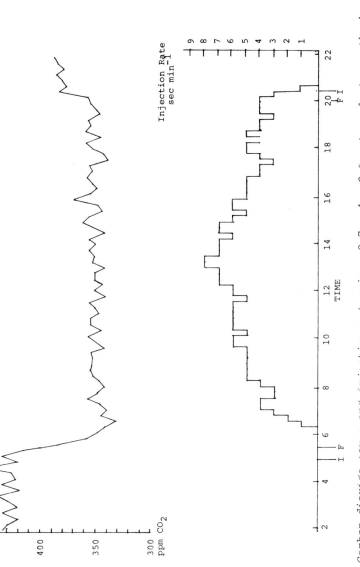

Figure 1. Carbon dioxide level and injection rates in a 2.7 x 4 x 2.6 meter plant growth chamber. The incandescent (I) lamps were turned on before and off after the fluorescent (F) lamps. Light intensity with all lamps lit is 625 microeinsteins m^{-2} s^{-1} PAR. The chamber contained 2 M tall Rottboellia exaltata (L.), itchgrass, C_4 plants.

ment and the handling of corrosive chemicals. The second method, and the one described by Pallas in the previous paper, involves the use of a second plant growth chamber. This second unit is filled with large photosynthesizing plants and the low CO_2 content air is exchanged with the test chamber air at the rate necessary to maintain the desired CO_2 level.

REFERENCES

Patterson, D. T., and Hite, J. L. (1975). A CO_2 monitoring and control system for plant growth chambers. *Ohio J. Sci. 75,* 190-193.

Downs, R. J., and Hellmers, H. (1975). "Environment and Experimental Control of Plant Growth." Academic Press, New York.

CARBON DIOXIDE: CRITIQUE II

Herschel H. Klueter

Agricultural Equipment Laboratory
USDA-SEA-Agricultural Research
Beltsville, Maryland

Several points should be made about Pallas' paper on carbon dioxide. First, there is a question of how a discussion of the global level of CO_2 really fits into this controlled-environment conference. Perhaps greater emphasis should have been given instead to reviewing the benefits of growing crops under increased levels of CO_2 or to a historical coverage of CO_2 use.

Second, a valid point was made about controlled environment studies and the lack of reporting of the CO_2 concentration. Until recently most researchers' studies have not been concerned with CO_2 content. Carbon dioxide is probably one of the least controlled factors in plant growth chambers. The type and size of the crop can have a tremendous effect on the rate of CO_2 depletion. The CO_2 level, in turn, affects the growth rate of the crop. The importance of external sources of CO_2 was also brought out. The most important source, of course, is human activity; both the number and proximity of humans can be significant. Cars, motors, and combustion processes all add to the background CO_2 level and, consequently, investigators should measure and control this parameter if repeatable studies are to be conducted.

235

Third, the discussion on measurement of CO_2 was rather brief.
Perhaps this topic could have been expanded. Six methods of
measuring CO_2 were mentioned. However, only two of these were
explained. At least a brief description should have been given
for each of the other techniques. With such a description,
scientists could decide for themselves whether they wanted to
use a particular method. It is true that the infrared absorption
technique is the principal method used; but with microprocessor
and microcomputer technology available, some of the other
techniques will probably develop. A discussion of CO_2 measuring
techniques was given by Bowman (1967).

The infrared gas analyzer can also be used as a differential
CO_2 analyzer. In this technique both the sample and reference
cells become active cells, and the gas flows through them. Calibration becomes somewhat more involved. The lowest desired level
of CO_2 gas, say 1000 ppm, is passed through both infrared
columns and the instrument is zeroed. Then, with this same gas
passing through the reference cell and a series of higher
concentrations of CO_2 passing through the sample cell, a calibration curve is developed for that instrument.

Klueter (1977) used 980 ppm as a "zero" gas, and a series
of gases above that value was used to develop the calibration
curve (Fig. 1). Once the curve was developed, only 1470 ppm of
gas were needed to calibrate the instrument. The calibration can
be done at any CO_2 level within the range of the instrument.

The fourth important point has to do with control of CO_2. A
nice presentation was made for a simple CO_2 control for a
single chamber. The requirements for control of CO_2 differ,
depending on the size of the chamber. For very small chambers
the sample of air drawn from the chamber comprises a significant
part of the total volume of the chamber, and the gas must be
returned to the chamber. Additionally, since the air sample
withdrawal reduces the pressure in the chamber, its return to the

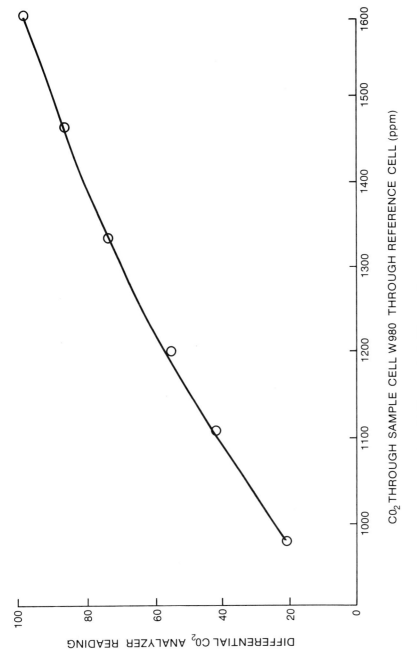

FIGURE 1. Calibration curve for differential CO_2 analyzer.

chamber restores the proper pressure. Use of this returning
gas line also serves as a convenient means of adding additional
CO_2.

For CO_2 control of several chambers with a single analyzer,
a pump and a three-way solenoid valve are required for each
chamber. The solenoid valve either returns the gas to the
chamber directly or diverts it through the analyzer. A complete
description of this type of system was given by Bailey et al.
(1970). When multiple chambers are being controlled by one
analyzer, the use of two lines to add CO_2 to each chamber is a
good technique. One line would be set by a needle valve to add
less CO_2 than is required in a continuous stream. The second
line would be controlled by the analyzer by means of a solenoid
valve and would add the remaining CO_2 intermittently to maintain
the desired level.

An unfortunate omission was that Pallas' paper did not discuss
the control of CO_2 in individual plant or leaf chambers. Such
control is especially important since the amount of CO_2 utilized
by the plant must be measured in many studies. A differential
analyzer previously described is required, and, if ambient CO_2
is used, no special requirements are necessary except to provide
a supply of air with a uniform concentration of CO_2. For
elevated CO_2 levels, Klueter (1977) added CO_2 to an air compressor
with a storage tank. A cylinder of CO_2 was placed near the
compressor with the pressure controlled line running to the
intake of the compressor. Also in the line was a solenoid
controlled by the compressor starter and a flow meter. When the
compressor started, the solenoid opened and concentrated CO_2 was
added to the air being compressed. A CO_2 level of about 1200 ppm
was maintained, but any level could be achieved by adjusting the
flow meter. When tests were being run, the compressor motor was
shut off so the concentration of CO_2 would not change. The CO_2
control system is shown in Fig. 2.

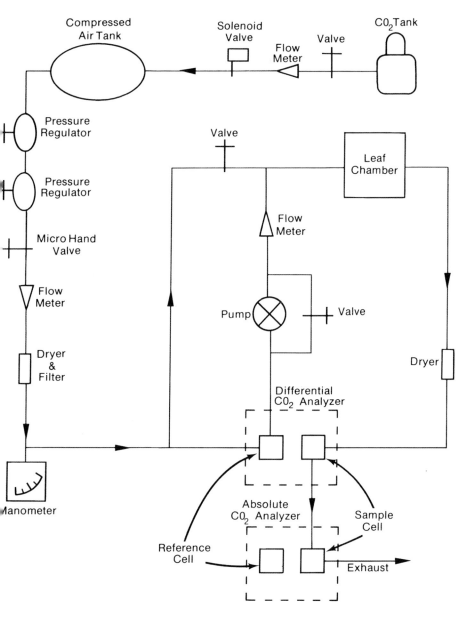

FIGURE 2. Flow diagram of CO_2 control system for the leaf chamber.

The next point has to do with units of measurements. Any discussion on measurement should address the subject of the units used. Units of microliters per liter ($\mu l\ l^{-1}$) were used by Pallas. We have used parts per million by volume (ppm). Others use volumes per million (vpm). Carbon dioxide used by the plants is generally measured in mg $dm^{-2}hr^{-1}$ (milligrams per square decimeter per hour) or cc $dm^{-2}hr^{-1}$ (cubic centimeters per square decimeter per hour). The question of units will be discussed more thoroughly in another part of this conference, but it does deserve at least a brief comment here.

Finally, a brief comment will be made about Pallas' Table 2. Three levels of CO_2 were tested and several parameters were measured. No information was given on *light*, *temperature*, *humidity*, or other factors. It is *very important* that all the major parameters be included in manuscripts. Many investigators have appropriately asked for such important information. It is hoped that these comments will help to move future reporting in that direction.

REFERENCES

Bailey, W. A., Klueter, H. H., Krizek, D. T., and Stuart, N. W. (1970). CO_2 systems for growing plants. *Trans. ASAE 13*, 263–268.

Bowman, G. E. (1967). The measurement of carbon dioxide concentration in the atmosphere. *In* "The Measurement of Environmental Factors in Terrestrial Ecology" (R. M. Wadsworth, ed.). Blackwell Scientific Pub., Oxford & Edinburgh, pp. 131–139.

Klueter, H. H. (1977). Comparison of pulsed and continuous light for photosynthesis with intact leaves. Ph.D. Thesis, Purdue University, Lafayette, Indiana.

CARBON DIOXIDE: GUIDELINES

Donald T. Krizek

Plant Stress Laboratory
USDA-SEA-Agricultural Research
Beltsville, Maryland

INTRODUCTION

Carbon dioxide (CO_2) concentration in the atmosphere has long been recognized as an important factor in photosynthesis (Smith, 1938; Decker, 1947, 1959; Gaastra, 1959, 1963; Moss, Musgrave, and Lemon, 1961; Loomis and Williams, 1963; Baker, 1965; El-Sharkaway and Hesketh, 1965; Hesketh, 1963, 1967; Brun and Cooper, 1967; Cooper and Brun, 1967; Ford and Thorne, 1967; Eastin *et al.*, 1969; Thorne, 1971; Zelitch, 1971; Chartier, 1972; Evans, 1975; Hand and Cockshull, 1975; Frydriych, 1976; Patterson, *et. al.*, 1977; Strain, 1978). Until the last 10 or so years, however, few investigators paid much attention to measuring or controlling CO_2 in plant growth chambers (French, Hiesey, and Milner, 1959; Krizek *et. al.*, 1968; Bailey *et. al.*, 1970; Zimmerman *et. al.*, 1970; Krizek *et. al.*, 1971; Krizek, 1974; Krizek *et. al.*, 1974; Patterson and Hite, 1975; Tibbitts and Krizek, 1978) or greenhouses (Lister, 1917; Gardner, 1964, 1965; Wittwer and Robb, 1964; Hand and Bowman, 1969; Wittwer, 1970, 1978; Pettibone *et. al.*, 1970; Kretchman and Howlett, 1970; Holley, 1970; Uchijima, 1971; Hand and Soffe, 1971; Rees *et. al.*, 1972; Slack and Calvert, 1972; Klougart, 1974; Enoch, Rylski, and Spigelman, 1976).

The objective of this paper is to briefly describe the guidelines and recommendations of the U.S. Department of Agriculture
(USDA), North Central Region (NCR)-101 Committee on Growth Chamber
Use for obtaining and reporting measurements of CO_2 concentration
in controlled-environment studies.

UNITS OF MEASUREMENT

Carbon dioxide concentration has traditionally been reported
in the United States as parts per million (ppm), and in the
United Kingdom and Europe as volume per million (vpm or ppmv).
A perusal of CO_2 literature published in the United States and
abroad reveals a considerable range in units reported (Table 1).
In some cases, authors use more than one unit of CO_2 concentration
in the same article (Badger, Kaplan, and Berry, 1977). Symposia
and monographs frequently contain several different units of CO_2
concentration (Burris and Black, 1976; Cooper, 1975). Standardization of the units for CO_2 concentration reported in the
literature would greatly facilitate making comparisons of experimental results.

The International System of Units

The International System of Units (or the Système International d'Unités) defines the SI units and prefixes. By international convention, seven physical quantities have been selected
for use as dimensionally independent base quantities: length, meter (m); mass, kilogram (kg); time, second (s); electric current,
ampere (A); thermodynamic temperature, degree Kelvin (K); luminous
intensity, candella (cd); and amount of substance, mole (mol) (The
Royal Society, 1971; National Bureau of Standards, 1977; Laidler,
1978). For each physical quantity, there is one and only one SI
unit. Decimal multiples of these units are constructed by the use
of 14 SI prefixes. The most common of these are: 10^6 mega (M);

TABLE 1. *Units of Carbon Dioxide Concentration Used in the Literature*

Units of CO_2 concentration	Author	Country	Year
ppm	Tibbitts and Krizek	U.S.	1978
ppm	Black et al.	U.S.	1976
ppm	Moss	U.S.	1976
v.p.m.	Monteith and Elston	U.K.	1971
v.p.m.	Bowman	U.K.	1968
$C(10^{-6})$	Acock, Thornley and Warren Wilson	U.K.	1971
μM	Schrader	U.S.	1976
μM	Hatch	Australia	1976
μM	Badger, Kaplan, and Berry	U.S.	1977
mM	Badger, Kaplan, and Berry	U.S.	1977
%	Cummings and Jones	U.S.	1918
%	Badger, Kaplan, and Berry	U.S.	1977
% in air	Zelitch	U.S.	1976
% Vol	Eliassen	U.S.	1974
% Vol	Madsen	Denmark	1976
mgm/l	Johansson	Sweden	1932
$\mu g/l$	Troughton	U.K.	1975
$\mu l/liter$	Quebedeaux and Hardy	U.S.	1976
$\mu l/liter$	Chang	U.S.	1975
μbar	Bjorkman, Badger, and Armond	U.S.	1978

10^3 kilo (k); 10^{-1} deci (d); 10^{-2} centi (c); 10^{-3} milli (m); 10^{-6} micro (μ); and 10^{-9} nano (n).

The concentration (or amount) or a substance based on the SI system is given in terms of moles per cubic meter, abbreviated as mol m^{-3} (National Bureau of Standards, 1977; The Royal Society, 1971). The mole (mol) was adopted as an SI base unit in October 1971 at a meeting of the Fourteenth General Conference on Weights and Measures (Mechtly, 1973). The mol is defined as the amount of substance of a system which contains as many elementary units as there are atoms in exactly 0.012 kilograms (or 12 grams) of the nuclide carbon-12 (^{12}C) (The Royal Society, 1971; National Bureau of Standards, 1977; Laidler, 1978). These units may be atoms, molecules, ions, radicals, electrons, photons or any other elementary particle or group of particles. Since 0.012 kg of carbon-12 contains the Avogadro number of atoms (6.02252 x 10^{23}), this is the number of elementary units contained in 1 mol (Morris, 1974; National Bureau of Standards, 1977). Thus the terms gram-equivalent, gram-molecule, and gram ions are obsolete and should be avoided (Morris, 1974). The SI unit for volume is the cubic meter (m^3), not liter as is commonly thought (Incoll, Long, and Ashmore, 1977; Laidler, 1978) although the term liter is still frequently used (Anon. 1978a). In order to adhere to the SI convention, only base units should be used for the denominator (Incoll, 1977).

Thus CO_2 concentration should be expressed as moles per cubic meter (mol m^{-3}) (National Bureau of Standards, 1977; Incoll, Long, and Ashmore 1977). Since this number may be cumbersome to work with, appropriate SI prefixes can be used in the numerator; for example, micromoles per cubic meter (μmol m^{-3}) or millimoles per cubic meter (mmol m^{-3}) (Incoll, Long, and Ashmore, 1977).

Adoption of the unit mol m^{-3} or its diminutive form (μmol m^{-3} or mmol m^{-3}) is consistent with recommendations made by the Crop Science Society of America (CSSA) Committee on Crop Terminology (Shibles, 1976) and by Incoll, Long, and Ashmore (1977).

CO_2 concentration may also be reported as micromoles per mole ($\mu mol\ m^{-1}$), which is numerically equivalent to ppmv, or cubic meters per cubic meter ($m^3\ m^{-3}$). However, these units are not as meaningful as the ones proposed.

CO_2 measurements obtained with a CO_2 infrared gas analyzer (IRGA) are recorded in terms of mass per unit volume while the CO_2 standards themselves are made up in terms of volume of pure CO_2 per unit volume of air or nitrogen. Regardless of the atmospheric pressure, the CO_2 standards will retain this ratio; however, the CO_2 concentration (in mass per unit volume or in moles per unit volume) as determined by the IRGA, will depend upon the temperature and pressure.

Because of this fact, it is recommended that CO_2 concentration be reported in $\mu mol\ m^{-3}$ or $mmol\ m^{-3}$ *at standard temperature and pressure (STP)* (where T at 0°C = 273.15 K; and P = 101,325 Pascals (Pa) (1.013 x 10^5 Newtons (N) m^{-2}). To convert ppmv CO_2 at STP to $\mu mol\ m^{-3}$ CO_2 at STP multiply by the conversion factor 44.6175*. To convert ppmv CO_2 at STP to $mmol\ m^{-3}$ CO_2 at STP multiply by the conversion factor 0.0446 (Table 2). A concentration of 1000 ppmn CO_2 would, therefore, be expressed as 44,617.5 $\mu mol\ m^{-3}$ or 44.6 $mmol\ m^{-3}$ at STP (Tables 2 and 3). Conversion factors for use at other temperatures from 0°C (273.15K) to 40°C (313.15 K) are given in Tables 4 and 5.

The conversion factor of 44.6175 is obtained by applying the equation for the ideal gas law PV = nRT (Morris, 1974) and solving for n where in SI units:

P = pressure of gas in Pa (at standard pressure = 101,325 Pa or 1.013 x 10^5 N m^{-2})

n = quantity of gas in mol

R = universal molar gas constant (at T of 0°C = 8.314 JK^{-1} mol^{-1})

The reader should note that the factor used by Incoll, Long, and Ashmore (1977) for converting ppmv to $\mu mol\ m^{-3}$ is incorrect by a factor of 100.

TABLE 2. *Conversion Factors for Expressing Carbon Dioxide (CO_2) Concentration in SI Units at Standard Temperature and Pressure (STP) (273.15 K and 101,325 Pa). To obtain CO_2 concentration in desired units, multiply the known unit in the left column by the conversion factor in the appropriate column of desired units.*

Known unit	Quantity	Desired units			
		$mmol\ m^{-3}$	$\mu mol\ m^{-3}$	ppmv	$mol\ mol^{-1}$
		Multiply by this factor			
$mmol\ m^{-3}$	(n_{CO_2}/v_{air})	–	1000	22.4127	22.4127
$\mu mol\ m^{-3}$	(n_{CO_2}/v_{air})	0.001	–	0.0241	0.0241
ppmv	(v_{CO_2}/v_{air})	0.0446	44.6175	–	1
$mol\ mol^{-1}$	(v_{CO_2}/v_{air})	0.0446	44.6175	1	–

$1\ mmol^{-3} = 1000\ \mu mol^{-3} = 22.41\ ppmv = 22.41\ mol\ mol^{-1} = 44.0\ mg\ m^{-1}$

T = temperature in K (T of $0^{\circ}C$ = 273.15 K)

V = volume of gas in m^3

Conversion factors (P/RT) for obtaining CO_2 concentration in $\mu mol\ m^{-3}$ or $mmol\ m^{-3}$ may also be determined for any given temperature and pressure by applying the equation for the Boyle-Charles law: concentration (at any T and P) = concentration (at STP) x (P/101,325) x 273.15/T; where P = pressure in Pa and T = temperature in K (Table 4). Conversion factors for changing CO_2 concentration from $\mu mol\ m^{-3}$ and $mmol\ m^{-3}$ to ppmv are given in Table 5.

By specifying the CO_2 concentration at STP one avoids the confusion of having the concentration of the standard gas vary with temperature and pressure. The problem of having the IRGA record CO_2 concentration at ambient temperature and pressure is

TABLE 3. Carbon dioxide (CO_2) Concentration Expressed in ppmv and μmol m^{-3} and mmol m^{-3} at Standard Temperature and Pressure (STP) (273.15 K, 101,325 Pa). Values in ppmv Multiplied by 44.6175 to obtain μmol m^{-3} and by 0.0446 to obtain mmol m^{-3} of CO_2 at STP.

CO_2 conc. ppmv	CO_2 conc. μmol m^{-3}	CO_2 conc. mmol m^{-3}
100	4461.75	4.46
200	8923.50	8.92
300	13385.25	13.38
400	17847.00	17.84
500	22308.75	22.30
600	26770.50	26.76
700	31232.25	31.22
800	35694.00	35.68
900	40155.75	40.14
1000	44617.50	44.60

TABLE 4. Conversion Factors for Changing Carbon Dioxide (CO_2) Concentration from ppmv to μmol m^{-3} and mmol m^{-3} at Different Temperatures and Standard Atmospheric Pressure (101,325 Pa).

Temp °C	Temp K	To obtain CO_2 concentration in	
		μmol m^{-3}	mmol m^{-3}
		Multiply ppmv by	
0	273.15	44.6175	0.0446
5	278.15	43.8155	0.0438
10	283.15	43.0417	0.0430
15	288.15	42.2949	0.0423
20	293.15	41.5735	0.0416
25	298.15	40.8763	0.0409
30	303.15	40.2012	0.0402
35	308.15	39.5498	0.0395
40	313.15	38.9183	0.0389

TABLE 5. Conversion Factors for Changing Carbon Dioxide (CO_2) Concentration from mol m^{-3} and μmol m^{-3} to ppmv at Different Temperatures and at Standard Atmospheric Pressure (101,325 Pa).

Temp °C	Temp K	To obtain CO_2 concentration in ppmv	
		From mmol m^{-3}	From μmol m^{-3}
		Multiply by	
0	273.15	22.4127	0.0224
5	278.15	22.3137	0.0223
10	283.15	23.2332	0.0232
15	288.15	23.6435	0.0236
20	293.15	24.0538	0.0241
25	298.15	24.4641	0.0245
30	303.15	24.8749	0.0249
35	308.15	25.2846	0.0253
40	313.15	25.6949	0.0257

partially overcome by the fact that it is a calibrated measurement and the calibration gas is specified at STP.

The chief advantage in reporting CO_2 on a concentration basis is that this is the variable sensed by the plant. The CO_2 flux is proportional to the difference in CO_2 concentration between the external environment and the sites of photosynthesis. Thus, anyone making photosynthetic measurements must at some point convert CO_2 exchange data to a concentration basis (Shibles, 1976; Incoll, Long, and Ashmore, 1977).

In view of the expressed philosophy of the USDA NCR-101 Committee on Growth Chamber Use to use SI units, it is recommended that CO_2 concentration be expressed as moles per cubic meter (mol m^{-3}) or its diminutive form μmol m^{-3} or mmol m^{-3} in future publications in the plant sciences and particularly for controlled-environment studies. The USDA NCR-101 Committee on Growth Chamber Use feels that the guidelines as distributed should be modified and made consistent with SI terminology and is prepared to modify them consistent with recommendations that develop out of the workshop. It would appear that the most appropriate unit would be either μmol m^{-3} or mmol m^{-3}.

CARBON DIOXIDE MEASUREMENTS

Where should CO_2 measurements be taken in the growth chamber? What kind of measurements should be obtained, and how should the data be reported? Bailey et al. (1970), Tibbitts and Krizek (1978), and others have described various methods for obtaining measurements of CO_2 concentration. The reader is referred to these sources for further information.

Assuming that a nondispersive infrared gas analyzer (IRGA) is used, it is recommended that CO_2 measurements be obtained at the mean canopy height level since at this location the boundary layer resistance is typically maximal. It is further recommended that CO_2 readings be obtained on a continuous basis, and that CO_2 concentration data for the course of a study be averaged on an hourly basis. Whenever possible, notes should be made on the CO_2 chart paper to explain inadvertent sources of CO_2 fluctuations in the growth chamber, (e.g. when the chamber door is opened, or if there has been an air pollution episode causing a high ambient level of CO_2). Since humidity interferes with the IRGA measurements, proper technical methods must be employed to prevent erroneous CO_2 values reported by the investigator. This can be accomplished by precise control of the chamber humidity or by utilizing inline water vapor condensing units to control the moisture in the sampled air arriving at the IRGA.

General air movement, PAR and total radiation levels, and CO_2-profiles above and within the plant canopy should be obtained to characterize the system. Hot wire anemometers, PAR sensors, and positioned IRGA sampling lines can be employed for this purpose.

CONCLUSIONS

Since the unit ppmv (or vpm) used in the literature is well
understood, the USDA NCR 101 Committee on Growth Chamber Use
feels that it is not necessary at this time to change units on
CO_2 measuring equipment (e.g., flow meters, calibration gases).
The Committee does, however, strongly recommend that the concentration of CO_2 reported be in SI units in order to standardize
reporting procedures.

A common scientific language should use SI units. What
better time to do so than now when other SI units are being
advocated? Converting from ppmv to $\mu mol\ m^{-3}$ (or $mmol\ m^{-3}$)
should be no more awkward than converting from bars to mega
Pascals or from Farenheit to Celsius. In order to obtain widespread adoption of the SI unit for CO_2 concentration and amounts
of other substances, it is recommended that the Council on
Biological Editors (1978) be urged to adopt SI units in their
publications. Editors of other plant science journals may
follow suit. Many journals have already made the shift to SI
units, e.g., the *Journal of Applied Ecology* (Anon. 1978a); the
Australian Journal of Plant Physiology (Anon. 1978b); and the
Biochemical Journal (Anon. 1978c). Other journals will inevitably
do so if given encouragement. Although the conversion to SI
units may be awkward initially, the long-term advantages outweigh
any short-term difficulties.

ACKNOWLEDGEMENTS

Grateful acknowledgments are extended to: Dr. Gaylon Campbell, Department of Agronomy and Soils, Washington State University, Pullman, Washington, and to Dr. Jesse Bennett, Plant Stress
Laboratory, U.S. Department of Agriculture, SEA, AR, for their
critical review of the manuscript and for their helpful sugges-

tions in preparing the revised draft; and to Mrs. Harriett Kilby, Plant Physiology Institute, SEA, AR, for her cheerful and helpful assistance in typing the manuscript.

REFERENCES

Acock, B., Thornley, J. H. M., and Warren Wilson, J. (1971). Photosynthesis and energy conversion. *In* "Potential Crop Production: A Case Study" (P. F. Wareing and J. F. Cooper, eds.), pp. 43-75. Heinemann Educational Books, London.

Anon. (1978a). A guide for contributors to the journals and symposia of the British Ecological Society. *J. Appl. Ecol.* *15*, 1-14.

Anon. (1978b). Notice to authors, Australian Journal of Plant Physiology. *Aust. J. Plant Physiol. 5(1),* 6 pp., not paged.

Anon. (1978c). Policy of the journal and instructions to authors. *Biochem. J. 169,* 1-27.

Badger, M. R., Kaplan, A., and Berry, J. A. (1977). The internal CO_2 pool of *Chlamydomonas reinhardtii:* response to external CO_2. Carnegie Inst. Washington Yearb. *76*, 362-366.

Bailey, W. A., Klueter, H. H., Krizek, D. T., and Stuart, N. W. (1970). CO_2 systems for growing plants. Proceedings Controlled Atmospheres for Plant Growth. *Trans. ASAE 13,* 263-268.

Baker, D. N. (1965). Effects of certain environmental factors on net assimilation in cotton. *Crop Sci. 5,* 53-56.

Bjorkman, O., Badger, M., and Armond, P. A. (1978). Thermal acclimation of photosynthesis: effect of growth temperature on photosynthetic characteristics and components of the photosynthetic apparatus in *Nerium oleander.* Carnegie Inst. Washington Yearb. *77*, 262-276.

Black, C. D., Goldstein, L. D., Ray, T. B., Kestler, D. B., and
Mayne, B. C. (1976). The relationship of plant metabolism to
internal leaf and cell morphology and to the efficiency of
CO_2 assimilation. *In* "CO_2 Metabolism and Plant Productivity"
(R. H. Burris and C. C. Black, eds.), pp. 113-139. University
Park Press, Baltimore.

Bowman, G. E. (1968). The measurement of carbon dioxide concentration in the atmosphere. *In* "The Measurement of Environmental Factors in Terrestrial Ecology" (R. M. Wadsworth, ed.),
British Society, Vol. 8, pp. 131-139. Blackwell Sci. Publ.,
Oxford.

Brun, W. A., and Cooper, R. L. (1967). Effect of light intensity
and carbon dioxide concentration on photosynthetic rate of
soybean. *Crop Sci.* *7*, 451-454.

Burris, R. H., and Black, C. C., eds. (1976). "CO_2 Metabolism
and Plant Productivity". University Park Press, Baltimore.

Chang, C. W. (1975). Carbon dioxide and senescence in cotton
plants. *Plant Physiol.* *55*, 515-519.

Chartier, P. (1972). Net assimilation of plants as influenced by
light and CO_2. *In* "Crop Processes in Controlled Environment"
(A. R. Rees, K. E. Cockshull, D. W. Hand, and R. G. Hurd,
eds.), pp. 203-216. Academic Press, New York.

Cooper, J. P., ed. (1975). "Photosynthesis and Productivity in
Different Environments". Cambridge University Press, Cambridge.

Cooper, R. L., and Brun, W. A. (1967). Response of soybeans to a
carbon dioxide-enriched atmosphere. *Crop Sci.* *7*, 455-457.

Council of Biology Editors (1978). "Council of Biology Editor's
Style Manual. A Guide for Authors, Editors, and Publishers
in the Biological Sciences", 4th ed. Am. Inst. Biol. Sci.,
Arlington, Virginia.

Cummings, M. B., and Jones, C. H. (1918). The aerial fertilization of plants with carbon dioxide. Vermont Agric. Expt.
Station Bull. 211.

Decker, J. P. (1947). The effect of air supply on apparent photosynthesis. *Plant Physiol.* *27*, 561-571.

Decker, J. P. (1959). Some effects of temperature and CO_2 on photosynthesis of *Mimulus*. *Plant Physiol.* *34*, 103-106.

Eastin, J. D., Haskins, F. A., Sullivan, C. Y., and van Bavel, C. H. M. (1969). "Physiological Aspects of Crop Yield." Amer. Soc. Agron. and Crop Sci. Soc., Madison, Wisconsin.

Eliassen, A. (1974). "Meteorology." *In* "Encyclopedia of Environmental Science" (D. N. Lapedes, ed.), pp. 316-322. McGraw-Hill, New York.

El-Sharkaway, M. A., and Hesketh, J. D. (1965). Photosynthesis among species in relation to the characteristics of leaf anatomy and CO_2 diffusion resistance. *Crop Sci.* *5*, 517-521.

Enoch, H. Z., Rylski, I., and Spigelman, M. (1976). CO_2 enrichment of strawberry and cucumber plants grown in unheated greenhouses in Israel. *Sci. Hort.* *5*, 33-41.

Evans, L. T. (1975). The physiological bases of crop yield. *In* "Crop Physiology: Some Case Histories" (L. T. Evans, ed.), pp. 327-355. Cambridge Univ. Press, London.

Ford. M. A., and Thorne, G. N. (1967). Effect of CO_2 concentration in growth of sugarbeet, barley, kale and maize. *Ann. Bot.* *31*, 629-644.

French, C. S., Hiesey, W. H., and Milner, H. W. (1959). Carbon dioxide control for plant growth chambers. Carnegie Inst. Washington Yearb. *58*, 352.

Frydriych, J. (1976). Photosynthetic characteristics of cucumber seedlings grown under two levels of carbon dioxide. *Photosynthetica 10*, 335-338.

Gaastra, P. (1959). Photosynthesis of crop plants as influenced by light, carbon dioxide, temperature and stomatal diffusion resistance. Meded. Landbouwhogesch. Wageningen. *59*, 1-68.

Gaastra, P. (1963). Climatic control of photosynthesis and respiration. *In* "Environmental Control of Plant Growth" (L. T. Evans, ed.), pp. 113-140. Academic Press, New York.

Gardner, R. (1964). CO_2 for glasshouse crops. (Great Britain) *Agriculture 71,* 204-208.

Gardner, R. (1965). The application of carbon dioxide enrichment to commercial grown glasshouse crops. *In* "Growers Annual and Research Digest", pp. 66-67. Grower Publ. Ltd., London.

Hand, D. W., and Bowman, G. E. (1969). Carbon dioxide assimilation and measurement in a controlled environment glasshouse. *J. Agr. Eng. Res. 14,* 92-99.

Hand, D. W., and Cockshull, K. E. (1975). The effects of CO_2 concentration on the canopy photosynthesis and winter bloom production of the glasshouse rose 'Sonia' (syn. 'Sweet Promise'). *Acta Hort. 51,* 243-252.

Hand, D. W., and Soffe, R. W. (1971). Light modulated temperature control and the response of greenhouse tomatoes to different CO_2 regimes. *J. Hort. 46,* 381-396.

Hatch, M. D. (1976). The C_4 pathway of photosynthesis: mechanism and function. *In* "CO_2 Metabolism and Plant Productivity" (R. H. Burris and C. C. Black, eds.), pp. 59-81. University Park Press, Baltimore.

Hesketh, J. D. (1963). Limitations to photosynthesis responsible for differences among species. *Crop Sci. 3,* 493-496.

Hesketh, J. D. (1967). Enhancement of photosynthetic CO_2 assimilation in the absence of oxygen as dependent upon species and temperature. *Planta 76,* 371-374.

Holley, W. D. (1970). CO_2 enrichment for flower production. *Trans. ASAE 13,* 257-258.

Incoll, L. D., Long. S. P., and Ashmore, M. R. (1977). SI units in publications in plant science. Commentaries in Plant Science, no. 28. *Current Adv. Plant Sci,* 331-343.

Johansson, N. (1932). A field experiment with the growth of sugar beets at different carbon dioxide content of the air. *Sven. Bot. Tidskr. 26,* 70-75.

Klougart, A. (1974). Integration of watering, sprinkling and
    CO_2 into the greenhouse programme. *Acta Hort.* *35*, 23-31.
Kretchman, D. W., and Howlett, F. S. (1970). CO_2 enrichment
    for vegetable production. *Trans. ASAE.* *13*, 252-256.
Krizek, D. T. (1974). Maximizing plant growth in controlled
    environments. *In* "Phytotronics III. Phytotronics in
    Agricultural and Horticultural Research." (P. Chouard and
    N. de Bilderling, eds.), pp. 6-13. Gauthier-Villars, Paris.
Krizek, D. T., Bailey, W. A., Klueter, H. H., and Cathey, H. M.
    (1968). Controlled environments for seedling production.
    Proc. Int. Plant Prop. Soc. 18, 273-280.
Krizek, D. T., Bailey, W. A., Klueter, H. H., and Liu, R. C.
    (1974). Maximizing growth of vegetable seedlings in
    controlled environments at elevated temperature, light and
    CO_2. *Acta Hort. 391*, 89-102.
Krizek, D. T., Zimmerman, R. H., Klueter, H. H., and Bailey, W. A.
    (1971). Growth of crabapple seedlings in controlled environ-
    ments: Effects of CO_2 level and time and duration of CO_2
    treatment. *J. Amer. Soc. Hort. Sci. 96*, 285-288.
Laidler, K. J. (1978). "Physical Chemistry with Biological
    Applications", pp. 556-562. Benjamin-Cummings, Menlo
    Park, California.
Lister, A. B. (1917). Carbon dioxide control of greenhouse air.
    Rept. Exp. Res. Sta. Chestnut, 1916, p. 9.
Loomis, R. S., and Williams, W. A. (1963). Maximum crop produc-
    tivity: an estimate. *Crop Sci. 3*, 67-72.
Madsen, E. (1976). Effect of CO_2 concentration on morphological,
    histological, cytological and physiological processes in
    tomato plants. Ph.D. Dissertation, Royal Veterinary and
    Agricultural University, Copenhagen. State Seed Testing
    Station, Lyngby, Denmark.

Mechtly, E. A. (1973). The international system of units.
    Physical constants and conversion factors. Second revision.
    Nat. Aeronaut. and Space Admin. SP-7012. Washington, D. C.
    21 pp. U.S. Govt. Printing Office, Washington, D. C.
Monteith, J. L., and Elston, J. F. (1971). Microclimatology and
    crop production. *In* "Potential Crop Production: A Case
    Study" (P. F. Wareing and J. F. Cooper, eds.), pp. 23-42.
    Heinemann, London.
Morris, J. (1974). SI units and their usage. *In* "A Biologist's
    Physical Chemistry", 2nd ed., pp. 14-25. Edward Arnold, London.
Moss, D. N. (1976). Studies on increasing photosynthesis in
    crop plants. *In* "CO_2 Metabolism and Plant Productivity"
    (R. H. Burris and C. C. Black, eds.), pp. 31-41. University
    Park Press, Baltimore.
Moss, D. N., Musgrave, R. B., and Lemon, E. R. (1961). Photo-
    synthesis under field conditions. III. Some effects of light,
    carbon dioxide, temperature, and soil moisture on photosyn-
    thesis, respiration, transpiration of corn. *Crop Sci. 1*, 83-87.
National Bureau of Standards (NBS) (1977). The International
    System of Units (SI). NBS Special Publication 330.
    Washington, D. C.
Nobel, P. S. (1974). "Introduction to Biophysical Plant Physiology.
    W. H. Freeman, San Francisco.
Patterson, D. T., Bunce, J. A., Alberte, R. S., and van Volkenburgh,
    E. (1977). Photosynthesis in relation to leaf characteristics
    of cotton from controlled and field environments. *Plant
    Physiol. 59*, 384-387.
Patterson, D. T., and Hite, H. L. (1975). A CO_2 monitoring and
    control system for plant growth chambers. *Ohio J. Sci. 75*,
    190-193.
Pettibone, C. A., Mason, W. R., Pfeiffer, C. L., and Ackley, W. B.
    (1970). The control and effects of supplemental carbon
    dioxide in air-supported plastic greenhouses. *Trans. ASAE
    13*, 259-262, 268.

Quebedeaux, B., and Hardy, R. W. F. (1976). Oxygen concentration: regulation of crop growth and productivity. *In* "CO_2 Metabolism and Plant Productivity" (R. H. Burris and C. C. Black, eds.), pp. 185-204. University Park Press, Baltimore.

Rees, A. R., Cockshull, K. E., Hand, D. W., and Hurd, R. G., eds. (1972). "Crop Processes in Controlled Environments." Academic Press, New York.

Schrader. L. E. (1976). CO_2 metabolism and productivity in C_3 plants: an assessment. *In* "CO_2 Metabolism and Plant Productivity" (R. H. Burris and C. C. Black, eds.), pp. 385-396. University Park Press, Baltimore.

Shibles, R. (1976). Crop Science Society of America (CSSA) Committee Report. Terminology pertaining to photosynthesis. *Crop Sci. 16,* 437-439.

Slack, G., and Calvert, A. (1972). Control of carbon dioxide concentration in glasshouses by the use of conductimetric controllers. *J. Agr. Eng. Res. 17,* 107-115.

Smith, E. L. (1938). Limiting factors in photosynthesis and carbon dioxide. *J. Gen. Physiol. 22,* 21-35.

Strain, B. R., ed. (1978). Report of the workshop on anticipated plant responses to global carbon dioxide enrichment. Duke Environmental Center, Duke University, Durham, North Carolina.

The Royal Society (1971). Quantities, units and symbols. A Report by The Symbols Committee of the Royal Society, London.

Thorne, G. N. (1971). Physiological factors limiting the yield of arable crops. *In* "Potential Crop Production: A Case Study" (P. F. Wareing and J. P. Cooper, eds.), pp. 143-158. Heinemann, London.

Tibbitts, T. W., and Krizek, D. T. (1978). Carbon dioxide. *In* "A Growth Chamber Manual: Environmental Control for Plants" (R. W. Langhans, ed.), pp. 80-100. Cornell Univ. Press, Ithaca, New York.

Troughton, J. H. (1975). Photosynthetic mechanisms in higher plants. *In* "Photosynthesis and Productivity in Different

Environments" (J. P. Cooper, ed.), pp. 357-391. Cambridge
University Press, Cambridge.

Uchijima, Z. (1971). The climate in growth chamber. Simulated
CO_2 environment and photosynthesis in a glasshouse. *Jap. J.
Agr. Meteorol.* 27, 45-57.

Wittwer, S. H. (1970). Aspects of CO_2 enrichment for crop
production. *Trans. ASAE 13,* 249-251.

Wittwer, S. H. (1978). Carbon dioxide fertilization of crop
plants. *In* "Crop Physiology" (U. S. Gupta, ed.), pp. 310-
333. Oxford and IBH Pub. Co., New Delhi.

Wittwer, S. H., and Robb, W. (1964). Carbon dioxide enrichment
of greenhouse atmospheres for food crop production. *Econ.
Bot. 18,* 34-56.

Zelitch, I. (1971). "Photosynthesis, Photorespiration and Plant
Productivity." Academic Press, New York.

Zelitch, I. (1976). Biochemical and genetic control of photo-
respiration. *In* "CO_2 Metabolism and Plant Productivity"
(R. H. Burris and C. C. Black, eds.), pp. 343-358. University
Park Press, Baltimore.

Zimmerman, R. H., Krizek, D. T., Klueter, H. H., and Bailey, W. A.
(1970). Growth of crabapple seedlings in controlled environments: Influence of seedling age and CO_2 content of the
atmosphere. *J. Amer. Soc. Hort. Sci. 95,* 323-325.

CARBON DIOXIDE: DISCUSSION

KRIZEK: I would like to mention a 1979 paper by Samish in *Photosynthetica* concerning some hazards of using desiccants in CO_2 systems. In addition to the problem of desiccants, several factors contribute to inaccurate CO_2 readings. Some tubing, for example, is very permeable to CO_2. As Pallas mentioned, moisture in the lines can also be a serious problem.

Large variations in CO_2 concentrations in chambers result from different amounts of laboratory activity over the weekend and during the week. Also on the east coast, there have been increases in CO_2 associated with significant air pollution episodes. It would be very helpful, therefore, if CO_2 levels could be monitored continuously, especially since the base-line CO_2 level may change. At Beltsville we recommended using 400 ppm of CO_2 as a base-line level because the CO_2 content of ambient air is often higher than 350 ppm. As the years go on, the base-line level probably will need to be raised. Pallas' data and our work shows that there is an enormous advantage in growing plants at elevated CO_2 levels; e.g., up to 500-600 ppm. In terms of CO_2 enrichment, the increase in CO_2 level to 1000 ppm will be the most cost effective. Increases above 1000 ppm will be beneficial but may be less cost effective.

LANGHANS: Hellmers, how do you go about isolating the chambers? If you have one chamber with 350 ppm CO_2 and another adjacent chamber that you want to run at 1000 ppm, how do you avoid keeping the CO_2 from intermixing?

HELLMERS: We really don't have any problems. The building is under positive pressure and we use a one pass air movement through the chamber area all of the time. Thus even though the chambers are close to each other the chambers themselves are relatively sealed. They do have some leakage but they don't leak that badly.

PALLAS: We have recently had problems with calibration gases. During the last year or so we've had two tanks that were off by as much as 40 ppm, based on our 4 or 5 other standard tanks. This empasizes that "calibrated" gases from commercial sources are not always reliable. We are now using a gas chromatograph for calibrating our CO_2 standards.

COYNE: We have experienced this problem during the past 10 years. We keep an inventory of standards that is referenced to Scripps Institute of Oceanography manometric standards.

The cylinder itself has a pronounced effect on concentration over time. Conventional cylinders commonly either release some CO_2 or absorb CO_2 over a period of time. However, we have not had problems when we specified new chrome-molybdenum steel cylinders baked out before filling. Aluminum cylinders are now available and they are particularly convenient for field use. They are much lighter than the chrome-molybdenum tanks. I don't know what their long term adsorption or desorption of CO_2 will be. We are still trying to evaluate this.·

We have also found considerable inaccuracy in the indicated specific activity of $^{14}CO_2$ labeled air in cylinders. Both the indicated CO_2 concentration and the specific activity of the CO_2 can vary significantly from the stated analysis. We have resorted to gas counting to check the specific activity and to compare CO_2 concentration against the Scripps standards.

I believe many of you are aware that the carrier gas for the CO_2 standard can influence the response of infrared gas analyzers. Equal concentrations of CO_2 in the air can give a different response than CO_2 in nitrogen because the O_2 in the air mixture has an influence on the response, which is not easily predictable. Another problem is that anlyzers from different manufacturers have different responses.

CURRY: I agree with Coyne about the reliability of various gas sources. I don't know of any really reliable sources. If

we could get some suggestions as to what the solution is I think
it would be helpful.

COYNE: John Kelly and I put together a report for an IBP
meeting. This gives procedures for obtaining high quality standards, how to prepare cylinders, and what to specify in requisi-
tions. I have a limited number of copies left and will send
them out on request.

KOSTKOWSKI: The experience at NBS confirms the comments already made about reliability of standard CO_2 in air mixtures from
various commercial sources. It also confirms the reliability of
the standards maintained at Scripps Institute. They compare to
NBS standards within about 0.5% of the CO_2 content and that is
quite good. A few years ago NBS had some CO_2 in air standards,
but there was so little demand for them that they were discontin-
ued. In a few months, new standards of CO_2 in nitrogen will be
available. However, the experts tell me that the infrared gas
analyzer, due to pressure broadening as was indicated, will give
a slightly different and inaccurate calibration but probably
only a few percent. So if the investigator is satisifed with 5%
accuracy, say 15 μl l^{-1} in ambient air, these new CO_2 in nitrogen
standards will be adequate and presumably more reliable than
those available commercially. If there is a broad enough demand
for CO_2 standards, I believe strong requests to NBS management
could make them available at a reasonable price, but a problem is
the priority of doing this and the intitial cost and effort.

PALLAS: I have a question concerning CO_2 concentration of
these standards. Do you know what these concentrations might be?
We need calibration gases of 300 and 600 μl l^{-1}, especially the
latter. We also need both CO_2 in nitrogen and and CO_2 in air.

KOSTKOWSKI: The standards will be CO_2 in nitrogen, but I
don't know the exact concentrations. There will be a range.
They are being developed for car exhaust determinations. Investigators can contact the appropriate person at NBS about this as
indicate in the CO_2 table in my paper.

PALLAS: I would encourage investigators to write to Ernest E.
Hughes and William D. Dorhow in the National Bureau of Standards
and spell out their requirement for CO_2 standards.

KOSTKOWSKI: The names you mentioned are those to contact
about technical aspects of standards in terms of getting a program for NBS to provide CO_2 standards in air in as short a time
as possible.   I should also mention that George Uriano, acting
Chief of the Office of Standard Reference Materials at NBS, ad-
ministers standards of this type and establishes priorities for
making standards available.

KRIZEK: For those making photosynthetic measurements, the
levels that Pallas suggested as CO_2 standards are appropriate.
For investigators doing CO_2 enrichment studies and particularly
when determining optimal levels, 1000 ppm, 2000 ppm, and perhaps
5000 ppm standards would also be needed.   The investigator can
also mix his own gases, but it is preferable to have a commercial
standard.

KOSTKOWSKI: I understand that CO_2 in nitrogen standards will
be available in about 3 months.   These will be 1% accuracy standards.   However, one must recognize that the total uncertainty
in measurements will be about 3% because another 1-2% uncertainty
must be added for the uncertainty of the infrared gas analyzer
itself.   If such accuracy is acceptable, the CO_2 standards that
will be available in a few months should be adequate.

PALLAS: Can we extrapolate CO_2 in nitrogen standards to CO_2
in air measurements?

KOSTKOWSKI: You cannot extrapolate.   This difference in reading between nitrogen and air will vary with the particular instrument you are using, so NBS cannot do this.   To make the correction for your instrument you would have to have a good standard.

PALLAS: Is it possible to obtain CO_2 standards in air?   This
is what nearly 99% of the investigators will need.   Only few
photorespiration studies are conducted with CO_2 in nitrogen.

KOSTKOWSKI: It should not be difficult to make CO_2 standards in air available to 1% uncertainty. The NBS investigators have a program more or less dictated to them so if you want this done within a year you would have to make a strong request, and show a real need in the United States for such a standard. I should add that there is a program for long-term climate modeling in which it is expected that a CO_2 standard in air with 0.1% uncertainty will be provided. This is at least ten times better than what you need. Such a CO_2 in air standard will be available in about 2 years while the CO_2 in nitrogen standard should be available in a few months. If investigators can demonstrate a significant need and a program is established, CO_2 in air standards with a 1% uncertainty could be made available in six months to a year. I understand that, since it is difficult to make a 0.1% CO_2 in air standard, it will only be in small supply in two years. However, it should be distributed to all the specialty gas manufacturers so that they should be able to provide you with reliable 1% CO_2 in air standards within a few years.

WALKER: For investigators using differential CO_2 analyzers the 1% uncertainty standards will not be very useful, so we certainly should not overestimate their use.

PALLAS: I would like to suggest that we vote to see if the group is in favor of encouraging NBS to develop CO_2 in air standards. If so, I recommend that we write a letter as a group to that effect rather than as individual investigators.

PRINCE: As I understand it, the CO_2 in nitrogen standard will be available in a short while. Will those of you who could use in less than 2 years CO_2 in air to 1% uncertainty raise your hands. *Note: more than half of the participants raised their hands.*

KOSTKOWSKI: If a letter is drafted to NBS, I would encourage you to try to make an estimate of the number of CO_2 standards that would be purchased per year by investigators interested in plant studies. As you will appreciate, there is continuous

pressure at NBS for setting up all types of standards for the
standard reference material program. I believe that if NBS has
an indication that 10 standards would be purchased per year and
someone else has written a letter indicating that 500 standards
will be needed per year, I think you know where the priorities
would go.

TIBBITTS: I will draft such a letter, please let me know the
standard concentrations you need and the number of cylinders of
each you plan to purchase.

McFARLANE: The price obviously will determine how much and
how often we would buy these cylinders. Is there a way of estimating cost?

COYNE: I will guess at the cost. I have recently purchased
standards in new cylinders filled by a commercial vendor and analyzed by Scripps Institute. The cost was approximately $650.00
per cylinder. One can get desired precision if he is willing to
pay for it.

TIBBITTS: I would like to ask Coyne to comment briefly on the
small CO_2 analyzer that his group is developing and which has
application in small growth chambers. Please indicate the source
of the analyzer if available and also the source of stainless
steel capillary tubing that you are using for slow metering of
concentrated CO_2 into chambers?

COYNE: We have been using capillary tubing to control CO_2
concentrations in cuvettes and porometers and will use the same
principle in our new growth chambers. Basically, we are working
with capillaries ranging in ID from 0.005 to 0.010 inch with
an OD of 1/16 inch. The tubing is stainless steel and compatible with compression fittings.[1] Tube length is varied to obtain
the desired flow rates for a given pressure range. We like to
work between 0 and 100 lbs pressure. An error signal from a CO_2

[1]Available from Tube Sales, 235 Tubeway St., Carol Illinois
60187.

sensor controls a motorized valve and either increases or decreases the pressure on the capillary tube. Since the flow is not
completely off, it is easier to timeshare a number of chambers on
the same analyzer and minimize the fluctuation in CO_2 concentration compared to an on-off system. The system Hellmers mentioned
is a marked improvement and would come close to the type of system we are using. Both systems are better than an on-off system.

Tibbitts' other question concerned development of a miniature
CO_2 sensor. My colleague, Gail Bingham, is the principal investigator of this project. We had a need for portable equipment for
measuring photosynthesis, transpiration, and stomatal conductance
in the field. We developed a mini-cuvette system that compressed
a van load of computer-controlled equipment into a hand-carried
box. With this instrumentation we can measure these plant pro-
cessses under controlled conditions of CO_2, temperature, water
vapor, and light intensity using a battery power source. The
mini-CO_2 sensor that we developed to be used with this equipment
has not yet been released. We do have working prototypes of a
sensor which employs a folded-path cell to attain a 48-cm path
length. This allows the sensor to be reduced to a size that can
be hand held. The instrument uses about 12 watts of power (12 v
DC) and it is lightweight and very sensitive (0.25 ppm at 320
ppm). It has a single source and detector and measures both CO_2
and water vapor concentrations by the ratio of nonabsorbing and
absorbing wavelengths. The unit will use a microprocessor to correct for temperature and pressure effects and to calculate concentration. A ratioing system is used to take care of any drift
that might occur as the mirrors become dirty.

McFARLANE: I would like to comment on the guidelines presented by Krizek. The infrared gas analyzers that I am familiar
with are not calibrated in ppm., but rather in units of 0 to 100.
It would be much easier for me to make that calibration chart in
the required units for reporting rather than converting to ppm
and then again to the necessary units for reporting.

KLUETER: I see one difficulty because all the gases are calibrated in ppm. If we can get the initial gases designated in
micromoles, or whatever, I think this would be fine.

HELLMERS: All one has to do is calculate the conversion of
ppm to the desired units for each cylinder and mark this on each
cylinder label.

KRIZEK: There is some precedent and strong rationale for ac-
cepting and backing SI units. The fact that I found 12 different units of CO_2 concentration in plant science research articles
makes for very difficult comparisons. Which is more cumbersome
for the investigator, to convert to $\mu mol\ m^{-3}$ or $mmol\ m^{-3}$ at the
time he is reporting his data, or for each of us to have to
familiarize ourselves with all of the required conversion
factors? I emphasize the paramount importance of using a common
language.

PALLAS: Who is going to abide by these recommendations? I
am a member of the American Society of Plant Physiologists and
I am not certain what units that organization will accept for
reporting. Will the horticultural journals accept what we recommend? And how will these recommendations be implemented?

HAMMER: I have discussed this problem with Janick, editor of
the Horticulture Society publications, and there appears to be
no question about recommendations being accepted as long as they
represent the consensus of the knowledgeable group within the
Society. The members of a society or a committee can have considerable influence on the editor. For example, in our case, the
ASHS Growth Chamber Committee published a number of recommendations on standard reporting procedures that were accepted by
the editor and members of our society. Therefore, investigators
may find it worthwhile to contact editors of their societies and
try to persuade them to accept our recommendations.

KRIZEK: One of the most important steps we could take would
be to forward the recommendations developed at this conference
to the Council of Biological Editors. A problem is that there is

little or no consensus for a coordinated effort to standardize reporting units among biological societies or biological journals. While individual societies may exert pressure on editors of journals to make such changes, we need some consensus among biological editors in general to do so. In England SI units for CO_2 concentration were accepted by the Faraday Society, the Royal Society, and several other societies that cut across the fields of biology, chemistry, physics, and medicine. Similarily in the United States we need to attempt to reach the audience of biological editors across the board. In my opinion, the Council of Biological Editors would be the most appropriate group to reach. If we cannot get general agreement we will not be successful in promoting wide adoption of our recommendations in individual journals.

PALLAS: I support Krizek. We have gone through some exercises in futility already and that is what happens when an investigator publishes in certain journals and doesn't know how he is supposed to report.

HAMMER: A question for Krizek. Shouldn't we add more significant figures to the conversion factors you presented? I have just made some conversions and if we try to interconvert 1000 ppm we need at least 4 significant figures to do that.

KRIZEK: That would certainly be the best approach.

KLUETER: I would like to ask Krizek if he had made any calculations on photosynthetic uptake of CO_2? What units would be used in describing photosynthesis according to the SI system instead of $mg\ dm^{-2}$ as commonly used?

KRIZEK: The unit for net CO_2 exchange would simply be $\mu mol\ m^{-2}\ s^{-1}$ or $mmol\ m^{-2}\ s^{-1}$. I refer you to the paper by Incoll (1978) and also to the Crop Science Society of America (CSSA) terminology report on photosynthesis (Shibles, 1976). It is quite straightforward to calculate CO_2 exchange in SI units if the CO_2 concentration is expressed in SI units. The important thing is that if you accept the SI convention, the base unit will

be meters rather than centimeters. Since the denominator is
supposed to be in terms of the base unit, the investigator can
change only the numerator, and he can do that by simply adjusting
the prefix. Thus, while the basic SI unit for CO_2 concentration
is mol m^{-3}, a less cumbersome unit to work with is the µmol m^{-3}
or mmol m^{-3}.

McCREE: I suppose the use of the mole for photosynthesis research is based on the idea that the CO_2 is converted into sugars
and that this is the end point, but of course this is a mistaken
assumption. Once we've adopted the mole of CO_2 fluxes, I agree
with Krizek that it would be consistent to use the mole for CO_2
concentration. I am not sure that this applies to controlled
environment studies in general. For those who do photosynthesis
research in controlled environments, this would be consistent.
The question was raised, if we specify CO_2 concentration in mol
m^{-3} why don't we do so for water vapor? The point is that there
is no chemical conversion - water is just flowing as a mass from
one point to another so that it is consistent to use mass units
for water flow and mole units for CO_2 flow. That's the rationale, I suppose.

FRANK: May I ask where CO_2 measurements will be taken and
what is the purpose of these measurements? I think Pallas said
that we need only be concerned about ± 15 ppm accuracy in making
CO_2 measurements. If we are simply characterizing the environment, this may be useful, but if we are looking at canopy photosynthetic levels in the chamber, are measurements at the top of
the canopy adequate? Should they be measured in the incoming
airstream? We are putting more and more over the top of the
canopy. In small, reach-in chambers we are reducing our space
considerably. If we inject CO_2 into a system has there been adequate mixing to obtain an accurate measurement over the canopy?
Air velocity should be considered. I don't believe mixing of
air is all that uniform in these chambers. Perhaps locating
the sensors in the air stream, after the stream has been

conditioned to the desired CO_2 level, humidity level, and temperature, is a better place to take measurements than actually over the canopy. I'd like to hear some comment about this.

KRIZEK: Even though an investigator is not measuring and controlling CO_2 concentration in a growth room or growth chamber, he should at least monitor and report it. That would be a great improvement over what we are now reporting. Large fluxes in CO_2 can occur in a growth chamber because of air pollution episodes, human activity, and the presence of mixed canopies of C3 and C4 plants. Hence it is essential that investigators have some data about CO_2 levels.

PALLAS: I think that along with reporting CO_2 concentration the investigator should be able to state that he calibrated his CO_2 analyzer according to NBS standards.

WENT: It was very interesting to see the curve of CO_2 utilization in the chamber Hellmers showed because it indicates how the rate of CO_2 utilization varies diurnally. Experiments in which the optimal CO_2 concentration has been established for different plants do not mean very much because, at different times of the day and under different physiological conditions, plants absorb CO_2 at appreciably different rates.

CONTROLLED ENVIRONMENT GUIDELINES FOR PLANT RESEARCH

WATERING

S. L. Rawlins

U. S. Salinity Laboratory
Riverside, California

INTRODUCTION

Experiments are usually conducted in controlled environments
to study the effects of environmental parameters on growth and
development of plants. Water plays such a crucial role in so
many plant processes that almost any parameter is influenced by
water stress if it is severe enough or long enough. For this
reason alone, one should not expect to study the independent effect of water on plants without coping with the highly integrated
set of physiological controls influenced by water. Add to this
the facts that water is the vehicle by which nutrients reach plant
roots and that it shares the pore space in the root medium with
an air phase through which oxygen is supplied. It is apparent
that correctly watering plants in controlled environments is crucial to conducting successful experiments.

Because it is impossible to treat by example all of the possible experimental situations one needs to be aware of to effectively water plants, this paper will first discuss the general
principles important in the processes of water retention and
transport in containerized growth media. With these principles,
individual cases can be solved by analysis. The remaining discussion will consist of a few important applications of these

271

principles involving watering plants, either to minimize water
stress or to control it as an independent variable.

PRINCIPLES OF WATER RETENTION AND FLOW

Water molecules are held within a porous medium primarily by
adhesive forces between them and the surface of solid particles.
The closer a water molecule is to a solid surface, the more
strongly it is held. For this reason, a given quantity of water
is held more tightly within a fine- than within a coarse-textured
porous medium. A measure of the energy with which a unit quantity
of water is held is the pressure potential. By convention, water
in a free standing pool outside the range of adhesive forces has
a pressure potential of zero. Because energy must be added to water retained in a porous medium to bring it to this standard
state, pressure potential of water held by adhesive forces in a
porous medium is always negative. Figure 1 shows the pressure
potential as a function of volumetric water content for porous
media of three different textures. Here the pressure potential
is expressed in terms of energy per unit weight of water, which
has the dimensions of length. This length, termed pressure head
(p), is the equivalent vertical height of a water column in equilibrium with the water in the porous medium. When p is negative,
it represents the length of a hanging water column that would be
in equilibrium with the medium, through the porous walls of a
tensiometer cup.

The water content at which $p = 0$ is approximately saturation;
that is, all pores are water-filled (some entrapped air practically always exists at $p = 0$). Below this water content, air oc-
cupies that pore space not filled with water. The volumetric air
content of the soil is, therefore, approximately equal to the
saturation water content minus the volumetric water content.

Water moves in a porous medium in response to both pressure
and gravitational heads. The gravitational head (z) is simply

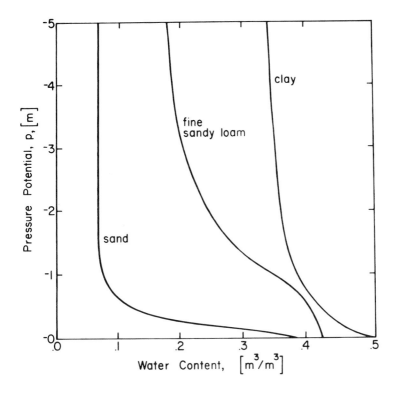

FIGURE 1. Water retentivity relationships for three soils.

the vertical distance between the point in question and an arbi-
trary reference plane. Water within a porous medium always tends
to move in a direction that decreases the sum of p + z, the hydraulic head, h. Thus for a uniform medium, water moves horizontally from wet to dry zones. It ceases to move horizontally when
the soil is uniformly wet; that is when p is constant throughout.
In a uniformly wet medium water moves downward in response to the
gravitational head until the water content of the upper zones
decreases to the point that p just balances z and h is the same
throughout. Because z decreases with depth with a gradient of
unity, p at equilibrium must increase with depth at the same gradient. The resulting water content distribution can be seen in
the water retentivity curves of Fig. 1. (The ordinate, in this

case, can be considered to be the container height.) Because
water moves in response to head gradients, water contents within
media of nonuniform textures will not be uniform. For example,
at p = -1 m, the sand and clay media shown in Fig. 1 will have
water contents of 0.10 and 0.38 m^3/m^3.

Because water is seldom at equilbirum within the soil, the
relationships obtained from Fig. 1 are primarily useful to indicate the direction in which water will tend to move within the
medium. In the dynamic case, where water is being added by irrigation, and lost by drainage, transpiration, and evaporation, the
rate of water movement becomes as important as the direction.
The rate of water movement is the product of the moving force and
the conductivity. The moving force for water flow is the gradient of the hydraulic head, grad h. (The gradient is the slope
of the curve of h vs. distance.) Therefore,

$$v = k \text{ grad } h$$

where v is the volumetric flux (volume per unit area per unit
time) and k is the hydraulic conductivity.

Water moves only through filled pores. It moves slowly near
soil particles and more rapidly further from the surface of particles. Consequently, the major factors governing hydraulic conductivity are water content and pore size distribution. Pore
size distribution depends on the size distribution of particles
making up the medium. Curves of hydraulic conductivity vs. water
content for the three media of Fig. 1 are given in Fig. 2. Within the pressure head range in which plants grow (0 to -150 m),
the hydraulic conductivity can vary by several orders of magnitude. This greatly affects the rate of water supply to plant
roots, and is a major cause for the difficulty in maintaining uniformly low values of p in a porous growth medium.

Proper interpretation of these few basic principles can explain why the answers to the fundamental questions of irrigation

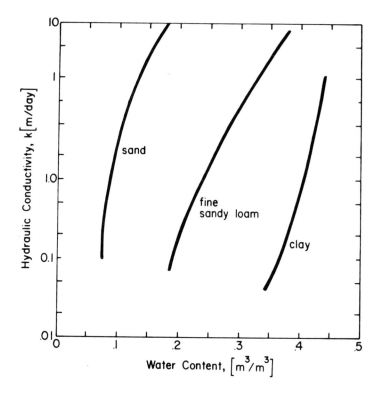

FIGURE 2. Hydraulic conductivity as a function of water content for three soils.

that is, when and how much water to apply may be different for containerized plants than for field-grown plants.

The frequency of irrigation required for a particular plant in a given environment depends on how much water can be stored within the rooting medium at each irrigation. It is generally understood that the quantity of water that can be stored depends on the volume of the container and the texture of the growth medium, but it is not so well understood why and how the quantity stored also depends on the shape of the container. Any water draining below the effective root depth is not available for crop use either in the field or in a container. In the field, the upper limit of available water, the "field capacity", describes the condition when downward flow beyond the root zone becomes

negligible. In many soil profiles, water movement does not suddenly change from significant to negligible. Field capacity is
rarely an equilibrium state, but rather occurs as a consequence
of a greatly diminished hydraulic conductivity within the profile
as the water content of the upper soil horizon decreases (see
Fig. 2). The force pulling water downward near field capacity is
usually dominated by a pressure head gradient arising from the
drier soil horizons beneath. For this reason, field capacity is
a function of the entire profile makeup, not merely that of the
surface horizons, and it is meaningless to speak of the "field
capacity" for a surface soil sample only.

At the bottom of a freely draining container, there are no
deeper horizons within which a pressure head gradient can exist
to pull water downward. Thus, for water to drain from the bottom of the container following a saturating irrigation, the pressure head must be zero. A water table (a zone of zero pressure
head), therefore, always exists at the bottom of any container
from which water is freely draining. At the time water ceases to
drain after an irrigation, the water-content profile within a
container will be that given in Fig. 1. The volumetric air content as a function of container height is approximately the difference between the water content at saturation and that given in
Fig. 1. Only with coarse-textured media and/or with deep containers, can waterlogging of an appreciable fraction of the soil
volume be avoided. For the same volume of porous medium, a deep
narrow container affords more air space per volume of soil than
does a shallow, wide container.

Porous suction tubes or plates can be used at the bottom of
each container to compensate for the absence of deep soil horizons to withdraw excess water. By controlling the suction within
these tubes, the lower boundary of the container can be maintained at a pressure head below zero. This simulates a water table
at a depth equal in magnitude to the negative pressure head maintained, permitting finer textured media to be used. By

withdrawing water from the lower part of the container by suction, the water content within the container can be brought to a more uniform and higher level without the presence of a waterlogged zone.

Aggregated porous media with dual porosities - that is, with large pores between the aggregates but small pores within them -- can also go a long way toward meeting the conflicting demands of adequate aeration and high water-holding capacity in containers. Coarse-textured organic materials are often used for this purpose. The large open pores drain quickly, but the porous aggregates hold considerable water within themselves.

Coarse-textured media without dual porosity, chosen to avoid waterlogging, require special irrigation practices. A system that delivers water at a low rate, such as some drip systems, is often unsatisfactory because it allows water to percolate through only a few open pores without wetting the entire medium. To a-void this, either the system must deliver water at a rate suffi-cient to flood the surface, or the drains must be temporarily closed or small enough to allow the container to fill.

Uniform water distribution and flow through the porous medium are particularly important if accumulation of soluble solutes within the growth medium is to be avoided. Although deionized water is commonly used for controlled environment experiments, added nutrients not taken up by the plant accumulate in the con-tainers. If the growth medium is not uniformly leached, pockets of high nutrient concentration can result. These subject the plant root system to non-uniform and uncontrolled nutrient levels. Particularly if the nutrients accumulate at the surface, merely changing the position of the water applicator can wash them into the active root zone, causing a salinity hazard. If tap water is used for irrigation, the salinity hazard is increased by any dis-solved solids.

The concentration of a given nutrient in a porous growth medium can either increase or decrease with depth, depending upon
the plant's uptake relative to the rate of supply. Two opposing
processes come into play. Plant uptake of a nutrient decreases
its concentration in the growth medium, while uptake of water reduces the quantity of solvent, increasing nutrient concentration.
An analysis based on the steadystate balance of mass gives the
relationship

$$\frac{C_o}{C_i} = \frac{1 - U}{L}$$

where C_o/C_i is the ratio of concentration of the nutrient in the
drainage to that in the irrigation water, U is the fraction of
the added nutrient taken up by the plant, and L, the leaching
fraction, is the ratio of the volume of drainage to irrigation
water.

Fig. 3 shows C_o/C_i as a function of U for three leaching fractions. On the line where $C_o/C_i = 1$, the concentration of the nutrient leaving the root zone is the same as that entering. It is
possible, therefore, to maintain an approximately constant nutrient concentration in a porous growth medium by balancing the rate
of supply of both water and nutrient. When $L > (1-U)$, then $C_o <
C_i$; and conversely, when $L < (1-U)$, then $C_o > C_i$. Thus a low
leaching fraction, when combined with nutrient supplies in excess
of plant uptake, can lead to high nutrient concentrations in the
lower root zone.

WATERING PLANTS TO ELIMINATE STRESS

It is doubtful that any technique can control the water, nutrient, and aeration status of plants as well as does nutrient
solution culture. This technique has three variations: (1) suspension of the root system in a solution-filled container into
which air is introduced; (2) suspension of the root system in an

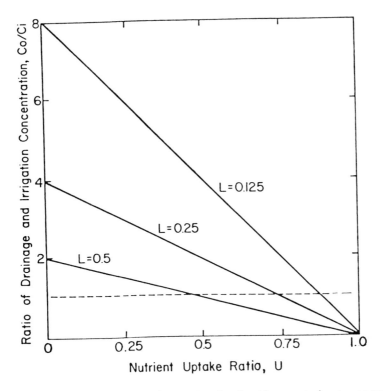

FIGURE 3. *Ratio of drainage to irrigation nutrient concentra-*
tions as a function of nutrient uptake ratio for three leaching
fractions.

air-filled container into which solution is introduced (as a
spray or an aerosol) under pressure; and (3) allowing the root
system to grow in a container of pure sand or gravel for support,
frequently filling it with nutrient solution to provide water and
draining it to provide aeration. A major advantage of all three
nutrient culture techniques is that they eliminate spatial gradients within the growth medium. This permits the entire root
system to be maintained at uniformly high levels of aeration,
nutrients, and water status.

SOLUTION-FILLED CONTAINERS

Significant improvements in techniques for manipulating nutrient solutions for research have been made since the modern introduction of solution culture by Hoagland and Arnon (1938). For
example, Gibson and Jolliffe (1972) devised a technique for automatically mixing batches of nutrient solution from manually prepared stock solutions, making available a continuous supply of
freshly mixed solution. The techniques are straightforward, using readily available electronics and electromechanical valves
and level detectors.

Even more sophisticated and useful for controlled experiments
is the system of Clement *et al*. (1974) that automatically monitors and maintains constant the activity of nutrient ions in a
flowing solution. By recording the amount of an ion required to
maintain a constant concentration, one can continuously monitor
nutrient uptake. Ion concentration in the flowing solution is
monitored by selective electrodes and individual nutrients are
added with chemical metering pumps. Transpiration can also be
measured by recording the quantity of deionized water required to
maintain the solution volume constant. Such a system has many advantages, but is not without drawbacks. Any system composed of a
number of electronic and electromechanical components is subject
to breakdown. Often this hazard will preclude the use of such
systems for long-term experiments.

If the nutrient solution is not automatically maintained at a
given concentration, the researcher must either replenish nutrients and water individually or replace the entire nutrient solution periodically. The frequency at which this is required depends on the capacity of the system, the rate of use, and range
of concentration variation that can be tolerated for each experiment. General rules of thumb cannot substitute for analyzing
the specific requirements of each experiment.

Tibbitts, Palzkill, and Frank (1979) describe an automated
system for continuously replacing water lost by evapotranspiration. Solution is pumped from a mixing chamber, through a

distribution manifold to individual plant growth containers, where it overflows and is returned to the mixing chamber through a collection manifold. A float switch in the mixing chamber opens a solenoid valve and admits new solution to replace that used or disposed of. A regulated quantity of solution is continuously disposed of through a capillary drain to maintain nutrient and H^+ concentrations.

Oxygen normally must be supplied to plants grown in nutrient-filled containers. Only with plants, such as rice or aquatic plants, can diffusion aerate the solution sufficiently for normal root growth. This must be accomplished by bubbling air directly into the solution in the plant growth container with non-recirculating systems. The bursting bubbles can cause problems by carrying nutrient salts to exposed plant leaves or by wetting foam plugs supporting the plant stem. Both processes can contribute to salt injury. Recirculating systems such as that described in Tibbitts, Palzkill, and Frank (1979) aerate the nutrient solution outside the plant grown container, avoiding this problem.

Mist Culture

A primary advantage of growing plants in mist rather than solution-filled containers is that the gas phase composition of the root environment can be independently controlled. Williamson (1968) described such a technique. Roots are grown in air-tight chambers where they are exposed to controlled mixtures of CO_2 and O_2. They are intermittently wetted with nutrient mist spray. Roots are normally sprayed for 1 minute out of 5.

Root systems of plants grown in mist culture typically are more branched and have greater root hair development than do roots of plants immersed in solution. Because no stored water is available if spray is not delivered, plants grown in mist culture are extremely vulnerable to mechanical or power failures that interrupt delivery of the nutrient mist. For this reason, mist culture will usually not be chosen for long-term or routine experiments, particularly if the root gaseous mixture does not require independent control.

Sand Culture

Sand (or gravel) culture is probably the simplest and most practical method for routine growth of plants in controlled environments. The sand serves two purposes: it supports the plant and stores nutrient solution for the plant between irrigations. If the sand is sufficiently coarse, the large air-filled pore space will be sufficient to provide adequate aeration after the water drains. The frequency of irrigation depends on the magnitude of this storage capacity relative to the rate of water use by the plant, as explained above.

The mechanics of filling and draining the containers can vary from manual to sophisticated automatic systems. Some degree of automation is usually desirable because the low storage capacity of the coarse sand makes it necessary to irrigate frequently. In any case, the irrigation system should completely fill the container, saturating the sand, before it is drained. To avoid aeration problems, the filling and draining process should comprise only a small part of the irrigation cycle.

Automated systems capable of filling and draining containers vary considerably in design and specific mechanical and electrical components. A primary consideration in choosing a system should be reliability. One of the simplest systems adaptable to containers whose tops are at the same elevation includes a standpipe connected to an outlet at the bottom of each container through a manifold. Periodically, the solution level in the standpipe is raised to a level corresponding to the top of the growth medium in the containers and then lowered below the bottom of the deepest container. It requires only a single conduit to each container for both filling and draining, and permits the

number and size of containers to be varied. The standpipe can be
filled by gravity through an electrical valve from an elevated
storage reservoir and then emptied by pumping water back to this
reservoir. The controls, however, can be simplified by using the
pump to fill the control standpipe from a lower storage reservoir,
and allowing the containers to drain by gravity. Electrical power is turned on by a clock timer to operate the pump and to close
an electrical valve in the drain of the level control reservoir.
When the water in the standpipe reaches the desired level, a float
switch turns off the power to the pump and the drain valve, allowing water to drain back to the reservoir. Once the power is in-
terrupted by the float switch, a reset relay keeps it off until it
is turned on again by the clock. It is important that the piping
to each container be large enough so that all containers fill at
the same rate.

A simple modification of this system uses a storage reservoir
located below the containers that can be pressurized. The clock
timer opens a 3-way electric valve introducing air pressure into
the reservoir and forcing water up into the containers. A float
switch in one container (or in a standpipe) disrupts power to the
electric valve, allowing the air to exhaust from the storage reservoir and the water to drain back. This eliminates the water
pump but requires a reliable source of air pressure.

With separate supply and drain pipes to each container, the
control system can be simplified by eliminating the float switch
and reset relay. We have used such a system (Hoffman and Rawlins,
1971) that pumped water into containers until it overflowed
through a tube just above the gravel growth medium. When the
pump was turned off, a small hole at the bottom of each container
allowed the solution to drain back to the reservoir. Similar systems have been designed that fill until a syphon drain is primed,
which then drains the container. With such systems, it is important that termination of filling be coordinated with the syphon-
ing sequence so that the container does not end up partly filled.

WATERING PLANTS TO MAINTAIN STRESS

In controlled environments, either the pressure or osmotic potential theoretically could be adjusted at the root surface to
control water stress. But, in an unsaturated growth medium, it is
virtually impossible to eliminate gradients of either of these
potentials. As the water content of the medium is allowed to decrease below saturation to lower the pressure potential from zero,
the hydraulic conductivity drops, consequently the gradient of p
required to maintain a constant water flux to the roots must increase proportionately. Because the hydraulic conductivity can
decrease several orders of magnitude within the growth range of
plants, gradients must be extremely high to maintain water uptake.
As a consequence, controlling p at some point in the growth medium
does not ensure controlling it at the root surface. Furthermore,
roots are never distributed uniformly, and water extraction rate
varies with position.

Two kinds of experiments have attempted to overcome the problem of a steep gradient of p in soil and maintain constant p at
the root surface. In the first the gradient is minimized by reducing the distance between the roots and the surface at which p
is controlled and the roots. Cox and Boersma (1967) attempted
this by growing plants in thin soil slabs that could be enclosed
in differentially permeable membranes. The seeds were germinated
and grown for a few weeks with normal irrigation from the top.
During short-term tests the membrane-enclosed slabs were immersed
in osmotic solutions to control p. The thin slab decreases, but,
of course, does not eliminate the gradient of p from the membrane
to the roots. If plants are grown continuously in such osmotic
cells, the roots tend to proliferate on the surface of the membrane. This decreases the gradient within the porous medium, but
does not eliminate it across the membrane. Without a direct measurement of p at the root surface, the actual value is always

subject to the uncertainty introduced by the limited permeability
of the membrane.

Hsieh, Gardner, and Campbell (1972) attempted to eliminate
this uncertainty by measuring water content, and inferring p, in
the vicinity of root hairs. They grew plants on a fine screen
that permitted only root hairs to penetrate into the soil beneath.
Water content in this root hair zone was measured with a collima-
ted gamma beam of a densitometer. The rate of water supplied to
the bottom of the soil column was then regulated to maintain a
constant water content in the root hair zone.

The severe limitations in root medium geometry of these experimental attempts to control p as an independent variable make them
impractical for any routine use. As a consequence we have no
practical means of maintaining p constant in an unsaturated porous
growth medium.

Attempts have been made to approximate the effects of constant
p by growing plants in solutions to which a non-absorbable solute
such as polyethylene glycol has been added (Lagerwerff, Ogata,
and Eagle, 1961; Janes, 1961; Jackson, 1962; Gee et al., 1973).
Presumably if the solute is not absorbed, the effect on the plant
is the same as though the water energy status were lowered by
force fields emanating from the surfaces of solid particles. To
the extent this presumption is valid, it is a satisfactory technique. At present, however, problems associated with toxicity or
uptake of contaminants and/or breakdown products of the solute
have not been eliminated. Kaufmann and Eckard (1971) found that
some polyethylene glycols sharply alter ion uptake characteristics of roots.

Considerable research has been performed using inorganic salts
to lower the osmotic potential of nutrient solutions. The effect
of these salts differs from that of the non-absorbable solutes.
Because the plant can take up absorbable salts and thereby adjust
its internal osmotic potential, there usually is no long-term effect on plant turgor. Caution should be taken to maintain

balanced ion ratios in nutrient solutions to avoid nutritional imbalances or toxicity.

Because the options available for controlling p in the root medium at a constant value are so limited, one must usually settle for a periodic cycling between upper and lower limits. The upper limit must always be high to raise the hydraulic conductivity to the point that the container can be wetted in a reasonable length of time. Water can be applied under a slightly negative pressure head as low as -0.8 m through a porous plate, but it is impractical to irrigate large containers at such low heads. The lower limit is determined by the pressure head at which irrigation is initiated. Because of the steep gradients of p within the soil, p measured by a soil sensor is not the same as that at the root surface. Particularly at low p, where the hydraulic conductivity is low, the difference in p between bulk soil and root surface can be extremely large. Numerous methods for measuring water content and water potential in soils are available. For advantages and disadvantages of each, see Rawlins (1976).

CONCLUSIONS

A wide variety of acceptable methods is available for growing plants at high water potentials. The most practical technique for routine environmental chamber research is undoubtedly some form of nutrient solution culture. Growing plants in containers of sand or gravel that are periodically filled with nutrient solution and drained appears to be one of the simplest ways of providing mechanical support, water, air, and nutrients to the root system. Stress levels for water, air, and nutrients can be minimized with an automatic nutrient solution culture. Monitoring plant water use with nutrient solution culture is simply a matter of recording the quantity of water required to restore the solution volume to its initial level.

The optimum stress level for a particular plant, however, is more difficult to determine. Although physiological responses of plants are beyond the scope of this paper, they certainly need to be considered in designing experiments to optimize the water regime while independently varying other parameters. For many plants, water stress may be necessary to inititate or promote reproduction or partitioning of photosynthate. For example, without some stress, onions fail to stop growing and do not translocate accumulated material from the leaves to form ripe bulbs. Sugar beets seem to require some stress to maximize synthesis of sugar.

Where some water stress is desired, it can be induced by adding to the nutrient solution solutes that may or may not be absorbed by the roots. Both absorbable and non-absorbable solutes induce decrease in plant turgor in the short term, but absorbable solutes allow turgor to be restored. One must be careful that secondary effects of the solutes do not overshadow the desired stress response, however.

It is generally impractical to control either the pressure or osmotic potential at the root surface at a fixed level in an unsaturated porous medium. A precipitous decrease in hydraulic conductivity with decreasing pressure head results in steep pressure head gradients in the medium. Because a uniform pressure head at the root surfaces in such a heterogeneous unstirred system does not exist, it is not possible to measure or control it. A uniform pressure head at the root surfaces exists only in a fully stirred system.

REFERENCES

Clement, C. R., Hopper, M. J., Canaway, R. J., and Jones, L. H. P. (1974). A system for measuring uptake of ions by plants from flowing solutions of controlled composition. *J. Exp. Bot.* *25*, 81-99.

Cox, L. M., and Boersma, L. (1967). Transpiration as a function of soil temperature and soil water stress. *Plant Physiol.* *42*, 550-556.

Gee, H. W., Liu, W., Olving, H., and Janes, B. E. (1973). Measurement and control of water potential in a soil-plant system. *Soil Sci. 115*, 336-342.

Gibson, J. S., and Jolliffe, P. A. (1972). A device for the automatic generation of plant nutrient solution. *Can. J. Plant. Sci. 52*, 409-412.

Hoagland, D. R., and Arnon, D. J., (1938). The Water-Culture Method for Growing Plants Without Soil. Univ. California Agr. Exp. Sta. Circ. 347, Berkeley, California.

Hoffman, G. J., and Rawlins, S. L. (1971). Water relations and growth of cotton as influenced by salinity and relative humidity. *Agron, J. 63*, 822-826.

Hsieh, J. J. C., Gardner, W. H., and Campbell, G. S. (1972). Experimental control of water content in the vicinity of root hairs. *Soil Sci. Soc. Amer. Proc. 36*, 418-420.

Jackson, W. T. (1962). Use of carbowaxes (polyethylene glycols) as osmotic agents. *Plant Physiol. 37*, 513-519.

Janes, B. E. (1961). Use of polyethylene glycol as a solvent to increase the osmotic pressure of nutrient solution in studies on the physiology of water in plants. *Plant Physiol. 36*, Suppl. xxiv.

Kaufmann, M. R., and Eckard, A. N. (1971). Evaluation of water stress control with polyethylene glycols by analysis of guttation. *Plant Physiol. 47*, 453-456.

Lagerwerff, J. V., Ogata, G., and Eagle, H. E. (1961). Control of osmotic pressure of culture solutions with polyethylene glycol. *Science 133*, 1486-1487.

Rawlins, S. L. (1976). Measurement of water content and state of water in soils. *In* "Water Deficits and Plant Growth". (T. T. Kozlowski, ed.) Vol. IV., pp. 1-55. Academic Press, New York.

Tibbitts, T. W., Palzkill, D. A., and Frank, H. M. (1979). Constructing a continuous circulation system for plant solution

culture. Res. Bul. R2963. Res. Div. Coll. Agr. and Life
Sci., Univ. Wisconsin, Madison, Wisconsin.

Williamson, R. E. (1968). Influence of gas mixture on cell division and root elongation of broad beans (*Vicia faba* L.).
Agron. J. 60, 317-321.

WATERING: CRITIQUE I

Merrill R. Kaufmann

Rocky Mountain Forest and Range Experiment Station
USDA-Forest Service
Fort Collins, Colorado

Water is so involved in the interaction of plants with their environment that physiological aspects of watering must be taken into account in any controlled environment study. This paper considers some of the points treated by Rawlins and examines several other physiological phenomena related to the water status of plants under controlled conditions.

Rawlins has described some of the physical aspects of supplying water to plants grown in containers for experimental purposes. Several of his points are significant physiologically and deserve emphasis. Poor drainage from containers after watering, which results in a saturated zone in the lower portion of the rooting medium, can have considerable significance in root aeration. Since roots are often proliferated near the bottom of the container, much of the root system may be exposed to a low-O_2 and high-CO_2 environment. When transpiration is low, as in the case of small plants or low light levels and high humidity conditions, saturation may persist for hours or even several days. Thus Rawlins' suggestion to use porous suction tubes or dual porosity rooting media to reduce the pressure head to some level below zero has considerable merit.

IRRIGATION FREQUENCY

The frequency at which irrigation is required for potted plants depends upon a number of factors. Obviously, the area of transpiring surface and volume of the potting medium occupied by

291

roots are important. Evaporative demand (e.g. vapor pressure or
absolute humidity difference from leaf to air) may be less important than once thought. Evidence is accumulating that
stomatal response to humidity results in reduced effects of
evaporative demand on transpiration over a rather broad range of
both natural and controlled conditions (Kaufmann, 1977; West and
Gaff, 1976). Humidity effects are considered in more detail by
Hoffman in this volume.

Guidelines for irrigation frequency also must depend on the
objectives of the experiment, however. If it is desired to keep
plants continuously well supplied with water, irrigation requirements and guidelines are necessarily very different than those
for experiments involving periodic soil drying. Rawlins notes
that, with techniques presently available, it is not possible to
maintain a truly uniform water supply (e.g. constant soil water
potential) at the surfaces of roots in a soil medium. Any extraction of water by the root system creates gradients for flow
of water in the soil and changes the soil water content at the
soil-root interface. Hydraulic conductivity of the soil changes
drastically with variation in water content, and uniform water
supply to the root is not possible over a range of root water
absorption rates and in the absence of a continual replenishment
of soil water.

Thus irrigation frequency becomes a matter determined by the
range of soil or plant water stress which can be tolerated in the
experiment. Most commonly, experimental designs call for
"adequate" or "optimal" water supply at all times; this is
assured by frequent irrigations whether they are needed or not.
Potential problems involving poor root aeration suggest caution,
however. A more desirable approach would be to give more
attention to the actual consumptive water use by the plant and
prevention of plant water stress. Slavik (1974) reviews many of
the methods potentially useful for monitoring plant water status.

Far too much emphasis is placed on so-called "optimum"

watering regimes for controlled environment experiments. Clearly
certain experiments dealing with physiological or morphological
aspects of plant growth are, by themselves, sufficiently complicated that avoiding plant water stress is desirable. It is
equally true, however, that plants grown under such conditions
generally differ physiologically and morphologically from plants
exposed to natural environmental conditions. Begg and Turner
(1976), in an extensive review of the subject, concluded that one
of the chief differences involves the response of plants to
water deficits.

One of the major effects of growing plants under continuously
well-watered conditions is that the plants assume physiological
and morphological characteristics typical for mesic sites, i.e.
plant tissues are not "hardened" by stress. It is well known
that plants grown under xeric conditions in the natural environment have a number of features not found in plants grown under
mesic conditions. Xeric conditions favor greater leaf thickness
and cuticular development, smaller cell size and leaf area, lower
shoot: root ratios, lower oxmotic potentials, and reduced cell
wall elasticity. From an experimental standpoint, the elucida-
tion of a biochemical pathway may not suffer if plant material
used for study is produced under growth environment conditions
quite different from those in the natural environment.

PLANT WATER RELATIONS IN THE FIELD AND IN CONTROLLED ENVIRONMENTS

But what about experiments in which growth or plant water
relations phenomena are studied in relation to particular environmental conditions? Begg and Turner (1976) provided numerous
examples showing that water relations characteristics differ for
plants grown in the field and in controlled environments; since
some of these factors have a direct relationship with growth
phenomena, growth responses also will differ, depending upon the

study environment. We have observed a number of "anomalous" plant responses under controlled environment conditions. For example, leaf
water potentials of sesame, sunflower, and pepper remained constant over a broad range of transpiration rates (Camacho-B, Hall
and Kaufmann, 1974). In other laboratories leaf water potential
of sunflower decreased as transpiration increased (Kaufmann, 1976).
The latter response would be expected under field conditions.
In another study, Camacho-B, Kaufmann, and Hall (1974) determined
the leaf water potential-transpiration relationship for citrus
seedlings grown under greenhouse conditions without significant
soil water deficits. For these seedlings, leaf water potential
was consistently higher (by roughly 5 bars) in the growth chamber
than observed in the field at equivalent transpiration rates.
Significantly, subjecting the seedlings to three successive drying
cycles in the growth chamber resulted in a water potential-transpiration relationship similar to that observed in the field.
Later studies by Ramos and Kaufmann (1979) showed that drying
cycles increased the hydraulic resistance of citrus roots compared with resistance determined on unstressed plants.

A number of factors, particularly radiation, temperature,
humidity, and watering, vary tremendously between field and controlled environmental conditions. The watering regime deserves
special attention in relating plant responses observed in controlled environment studies to field responses. Recalling
Rawlins' comment that "water plays such a crucial role in so
many plant processes", it seems appropriate to reevaluate the
practice of watering to maintain so-called "optimum" soil water
supply in many types of controlled environment studies. This is
a major limitation in growth chamber research at the present time.

CONTROLLING SOIL WATER STRESS

The use of drying cycles is perhaps the only appropriate
method for simulating the natural stress conditions of the field

environment. Other methods have been used in attempts to maintain constant reduced levels of water supply to roots (see
Rawlins' paper, this volume). For most experimental purposes,
however, attaining nearly constant levels of soil or root water
stress is as atypical as maintaining continuously well-watered
conditions. The routine use of drying cycles in controlled experiments has been limited by the difficulty and time required
for evaluating the degree of stress imposed. Unless root systems
are confined to small soil volumes, roots are unevenly distributed in the soil mass. Consequently, under conditions resulting
in high transpiration rates, water is absorbed primarily from the
portion of the soil container having the highest density of roots.
Under low transpiration conditions (e.g. during the night), however, most water may be absorbed from the wetter soil zones having
only a few roots. Thus plant water potential during the daytime
may reflect the soil water potential in the zone of maximum root
density (the driest part of the soil), whereas at night plant
water potential may reflect the highest soil water potential in
the rooting medium. J. A. Adams (unpublished) observed such a
response in *Simmondsia* early in a drying cycle. While this pattern of response complicates the assessment of soil and plant
water stress for potted plants under controlled conditions, the
subject should not be ignored because similar responses occur in
the field. Under orchard conditions, Sterne, Kaufmann and Zentmyer
(1977) found that depletion of water in the upper 90 cm of soil
reduced daytime leaf water potentials of avocado trees several
days before effects on nighttime potentials were observed, apparently because water remained adequately available to those
roots extending below 90 cm. It appears, therefore, that subjecting potted plants to drying cycles provides a reasonably valid
simulation of natural phenomena. The use of plant water stress
measurements (e.g. pressure chamber) is probably the most appropriate approach for assessing the severity of drying cycles, even
though such measurements may not indicate the dynamic nature of the

absorption process in the rooting zone.

STOMATAL CYCLING

Periodic oscillations of stomatal aperture with periods of one-half to 2 hours have been reported numerous times. Levy and Kaufmann (1976) showed that stomatal cycling can occur in mature citrus trees under field conditions. Unfortunately for the investigator who uses controlled environments, however, almost every other report of stomatal cycling comes from studies using controlled growth conditions. The true extent of cycling under either controlled or natural conditions is not known. Cycling is detected by measuring stomatal conductance, leaf-to-air temperature difference, leaf water potential, or tissue thickness at intervals of every several minutes or less. Most studies either do not include such measurements or the frequency of observation is so low that cycling is not detected. Thus it is not known how widespread the phenomenon is.

Stomatal cycling can have such a pervasive influence on controlled environment experiments that its effects cannot be ignored. Imagine an investigator attempting to measure transpiration based on hourly changes in pot weight when leaf conductance is cycling from 0.40 to 0.03 cm s^{-1} with a period of 40 minutes. Or suppose a study is conducted to determine the effects of temperature on growth when one of the treatments induces cycling. Undoubtedly many of the 100 or more reports of cycling resulted when an experiment was undertaken for a very different reason, only to have cycling obliterate the chances for a sound interpretation of the original hypothesis.

Apparently cycling can be caused in a number of ways. With citrus seedlings, we have induced cycling most frequently (but not always) by rapid changes in environmental conditions. For example, at constant light intensity, a 10°C shift in temperature regime or a 30% change in relative humidity often resulted in

stomatal oscillations which sometimes lasted for the rest of the
light period. Cycling has been observed during moderate water
stress, and it also has followed irrigation. Even simply opening
the door of a reach-in growth chamber to examine or move plants
may induce oscillations. Furthermore, once plants exhibit
cycling, it is often difficult to prevent it on successive days
by altering environmental conditions.

Attempts have been made to determine the exact nature of
cycling from a physical and physiological point of view (Cowan,
1977; Farquhar and Cowan, 1974). Such studies have proved difficult because of the wide range of conditions which result in
cycling. Nonetheless, a thorough understanding of the nature of
stomatal oscillations is badly needed to learn if certain procedures may be followed to avoid this complication in controlled
environment research.

In conclusion, several physiological considerations must be
taken into account in managing irrigation practices and plant
water relations during controlled environment studies. First,
attention should be given to achieving a reasonable balance between consumptive water use and irrigation frequency. Proper
selection of the container apparatus and rooting medium along
with suitable scheduling of irrigation can minimize soil saturation and keep salt accumulation acceptably low. Secondly, the
so-called "optimum" supply of water (i.e. frequent irrigation)
may not be optimal at all if experiments involve relating growth
or water relations phenomena to environmental conditions. Drying
cycles provide one means of making controlled environment experiments simulate more closely the natural environment. Finally,
stomatal cycling can be induced readily in controlled environments. Since cycling can drastically alter plant-environment
interactions, plants should be monitored to learn if the phenomenon exists during an experiment.

REFERENCES

Begg, J. E., and Turner, N. C. (1976). Crop water deficits. *Adv. Agron.* 28,161-217.

Camacho-B, S. E., Hall, A. E., and Kaufmann, M. R. (1974). Efficiency and regulation of water transport in some woody and herbaceous species. *Plant Physiol.* 54,169-172.

Camacho-B, S. E., Kaufmann, M. R., and Hall, A. E. (1974). Leaf water potential response to transpiration in citrus. *Physiol. Plant.* 31,101-105.

Cowan, I. R. (1977). Stomatal behavior and environment. *Adv. Bot. Res.* 4,117-228.

Farquhar, G. D., and Cowan, I. R. (1974). Oscillations in stomatal conductance - the influence of environmental gain. *Plant Physiol.* 54,769-772.

Kaufmann, M. R. (1976). Water transport through plants: current perspectives. *In* "Transport and Transfer Processes in Plants" (I. F. Wardlaw and J. B. Passioura, eds.), pp. 313-327. Academic Press, New York.

Kaufmann, M. R. (1977). Citrus - a case study of environmental effects on plant water relations. *Proc. Int. Soc. Citriculture* 1,57-62.

Levy, Y. and Kaufmann, M. R. (1976). Cycling of leaf conductance in citrus exposed to natural and controlled environments. *Can. J. Bot.* 54,2215-2218.

Ramos, C., and Kaufmann, M. R. (1979). Hydraulic resistance of rough lemon roots. *Physiol. Plant.* 45,311-314.

Slavik, B. (1974). "Methods of Studying Plant Water Relations." Springer-Verlag, New York.

Sterne, R. E., Kaufmann, M. R., and Zentmyer, G. A. (1977). Environmental effects on transpiration and leaf water potential in avocado. *Physiol. Plant.* 41,1-6.

West, D. W., and Gaff, D. F. (1976). The effect of leaf water
    potential, leaf temperature, and light intensity on leaf
    diffusion resistance and the transpiration of *Malus syl-
    vestris*. *Physiol. Plant.* 38,98-104.

WATERING - CRITIQUE II

G. S. Campbell

Department of Agronomy and Soils
Washington State University
Pullman, Washington

INTRODUCTION

Rawlins' paper discussed the principles that apply to watering and suggested methods to assure that plants receive an adequate supply of water and nutrients throughout an experiment. Rawlins also discussed some of the methods that have been used to control soil water potential in controlled environment situations. This paper will consider watering procedures to achieve a given level of moisture stress. Measurement of the water status of plants and soils will also be discussed.

In general, controlled environment studies which have water supply as a variable belong to one of two groups. In the first group are investigations to determine plant response to a particular water stress condition. The second group includes studies that are intended to simulate water stress under natural conditions. Examples of experiments in the first group may be those in which stress is imposed by severing the plant from its roots (or a leaf from a stem), adding solutes to the nutrient solution or soil, or controlling the water potential of soil in which roots are growing. Each experiment of this type has the objective of determining how a plant, grown under a specified set of conditions, will respond to water stress imposed in a particular way.

301

Such experiments are useful in gaining understanding of basic physical and physiological principles if the imposed stress and the plant's response are carefully monitored and reported.

Examples of the second group include experiments designed to study effects of water stress in controlled environments in such a way that the results of the experiments can be extrapolated to field situations. In such experiments it is important to match both plant water supply and plant water demand as closely as possible to conditions that prevail when the plants grow in the field. It is also important to compare osmotic potentials, turgor pressures, stomatal resistances, elongation rates, etc. between field and controlled environment plants to assure that the imposed treatment does result in the stress levels which exist in the field.

WATER RELATIONS IN CONTROLLED AND FIELD ENVIRONMENTS

The water related parameters which appear most closely associated with plant growth and development are osmotic potential (OP) of the cell sap, turgor pressure of the cells, and the stomatal diffusion resistance.

Work from a number of sources shows significant differences between osmotic potentials of well-watered field and controlled environment plants. For example, Campbell et al. (1976) found OP's of -8 bars for growth chamber grown potatoes, compared to -10 bars for field grown plants. Similarly, OP's of growth chamber wheat averaged -15 bars compared to -21 bars for field grown wheat. Papendick and Campbell (1974) attribute such differences to variations in photosynthetically useful light in different environments. As evidence of this they cited results of an experiment in which OP's of field grown wheat increased (became less negative) by 2-4 bars within a week following shading with 50% shade cloth.

The turgor pressure in plant cells is equal to the difference between water potential and osmotic potential in the cells. The plant water potential is determined by transpiration rate, resistance to water flow in the soil-plant system, and soil water potential. Since expansion growth is extremely sensitive to small changes in turgor below some threshold value (Hsiao, 1973; Boyer 1968), osmotic potential and water potential are important in determining rate of growth.

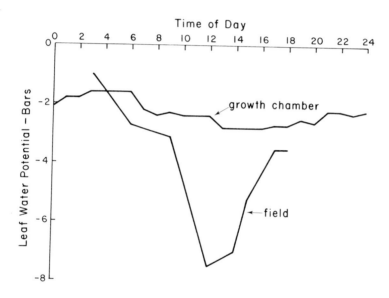

FIGURE 1. Comparison of leaf water potentials in field and growth chamber potatoes. (Data from Campbell et al., 1976).

Fig. 1 compares a typical diurnal leaf water potential cyle for potatoes in a growth chamber with similar measurements from the field (Campbell et al., 1976). Osmotic potentials for the two environments were -8 and -10 bars, as previously mentioned. These data indicate substantially lower daytime water potentials and at times higher night time potentials in the field than exist in the growth chamber. The growth chamber in which these

measurements were made had a 14 hr light cycle at a temperature
of 30°C and a 10 hr dark cycle at a temperature of 22°C. Light
was from fluorescent tubes having illuminance of 11,000 to 15,000
lux at the soil surface. Plants were grown in containers having
dimensions of 30 X 30 X 32 cm. The average daily transpiration
rate (per unit soil surface area) was 7.6 mm day^{-1} compared to
9 mm day^{-1} for the field grown potatoes, but peak rate in the
growth chamber was only 13 mm day^{-1} compared with peak rates >20
mm day^{-1} in the field. In the growth chamber, transpiration decreased when leaf water potentials dropped to around -3.5 bars
whereas in the field, stomata did not close measurably, even at
leaf water potentials of -8 bars.

In another study Papendick and Cook (1974) found that stress
under field conditions was highly correlated with incidence of
Fusarium foot rot in wheat. Glasshouse experiments (R. J. Cook,
personal communication) were conducted to simulate field water
stress conditions, but were unsuccessful in producing foot rot in
infected plants even though water was withheld until plants wil-
ted. Susceptible plants lowered osmotic potentials as much as 20
bars in the field (Papendick and Cook, 1974) in response to slowly imposed stress. This osmotic adjustment was not achieved in
the glasshouse, even when plants were severely stressed. The low
osmotic potentials attained in the field apparently were important in predisposing plants to foot rot attack, and these low potentials could not be induced with conventional controlled environment stressing procedures. In a later study (Papendick and
Campbell, 1974) osmotic potentials comparable to those measured
in the field were obtained under glasshouse conditions by plant-
ing wheat in tubes 150 cm long and then withholding water so that
water became gradually less available as the zone of extraction
moved deeper into the soil.

These examples are not meant to indicate that the field environment cannot be simulated by a controlled environment. They do
show, however, that the water relations of typical controlled

environment plants differ from those of field-grown plants. Some
reasons for these differences will now be considered.

DESIGN OF WATER STRESS TREATMENTS FOR CONTROLLED ENVIRONMENTS

To simulate the field moisture regime the following requirements must be met: (1) light levels must be sufficiently high for
photosynthesis to proceed at rates comparable to those under field
conditions, (2) potential transpiration rates must be comparable
to those in the field, (3) resistances to water transport from the
soil to the leaf must be similar to field resistances, and (4) water potential distributions in the root zone must be similar to
those found in the field.

Measurement and control of light are discussed elsewhere in
this volume. To determine whether photon flux is adequate for developing osmotic potentials comparable to levels found in the
field, the investigator needs only to compare osmotic potentials
under both conditions (methods for measurement will be discussed
later).

The transpiration rate of a leaf depends on leaf temperature,
atmospheric vapor concentration, and resistance to vapor transport from the substomatal cavities to the external environment.
Leaf transpiration rate, E_ℓ (g m^{-2}s^{-1}) can be expressed as (Campbell, 1977):

$$E_\ell = (\rho'_{vs} - \rho_{va})/r_v \tag{1}$$

where ρ_{vs} and ρ_{va} are saturation vapor density (gm^{-3}) at leaf
temperature and air vapor density, respectively, and r_v is the total resistance (sm^{-1}) to vapor diffusion (sum of stomatal and
boundary layer resistances). Since ρ'_{vs} is a function of leaf
temperature, which, in turn, depends on radiant energy load, convective cooling, and transpiration rate, increasing wind speed in
the leaf environment may either increase or decrease E_ℓ.

For plants growing in small containers (typical of controlled environment experiments) the transpiration rate (E_p) per unit of surface area is:

$$E_p = (\rho'_{vs} - \rho_{va})/\bar{r}_v \tag{2}$$

where ρ'_{vs} is the mean saturation vapor density of the leaves and $\bar{r}$ is the equivalent parallel resistance of all leaves on the plant, each weighted by the appropriate leaf area. If all leaves have the same r_v, then the transpiration rate per unit of soil surface area is

$$E_p = LAI \cdot E\ell \tag{3}$$

where LAI is the leaf area index for a potted plant (leaf area per unit soil surface area). Transpiration rate computed on a unit soil area basis is the most appropriate measurement for comparison to field transpiration rates because it relates to both atmospheric demand and water supply capability of the soil. Equation 3 indicates that E_p can be adjusted either by adjusting atmospheric vapor concentration or leaf area index.

It will usually not be practical to measure all of the variables necessary to compute E_p from eq. 2, though rough computations using typical or expected values of the parameters may be useful in designing an experiment. A more direct method for finding E_p is to measure transpiration rates of representative plants by weighing the pots. A typical field value for E_p in arid areas is 7 kg m^{-2} day^{-1}, and values twice that high have been measured in the presence of strong advection (Rosenberg and Verma, 1978). In humid areas, typical values are 2-3 kg m^{-2} day^{-1}. The peak value of E_p at mid-day is typically 2 to 2.5 times the mean value, and E_p is at or near zero at night because $\rho_{va} \simeq \rho'_{vs}$ and r_v is large from stomatal closure. Controlled environments may fail to simulate the $\rho_{va} \simeq \rho'_{vs}$ condition at night, so water potentials often do not recover to values near zero, as is typical in the field (Fig. 1).

Over the past several years considerable effort has gone into development of systems for control of water potential in the root zone of a plant at some specified level. These have been useful for determining plant responses to stress and magnitudes of resistances to water flow, but are not relevant if an investigator wishes to simulate field water stress conditions. In the field, part or all of the root zone is recharged to field capacity each time water is applied, and stress is imposed as the plant dries the soil. Simulation of field conditions requires that rooting depth and density, soil hydraulic properties, water storage, and temperature of the rooting medium be sufficiently similar to those existing in the field. If we consider water transport through the plant in more detail, the importance of each of these factors will become more apparent.

Following van den Honert (1948), transpiration (E_p) of the plant is directly proportional to leaf water potential (Ψ_ℓ) minus soil water potential (Ψ_s), and inversely proportional to the sum of the resistances (R) to water flow in the soil-plant system:

$$E_p = \frac{-(\Psi_\ell - \Psi_s)}{R_s + R_r + R_x + R_\ell} = \frac{-(\Psi_\ell - \Psi_s)}{R_t} \qquad (4)$$

where the subscripts on the resistance indicate resistance to water flow in the soil (s), root endodermis (r), xylem (x), leaf (ℓ), and total resistance (t). Rearranging the second part of eq. 4, it may be seen that $\Psi_\ell = \Psi_s - R_t E_p$, indicating that the amplitude of diurnal leaf water potential fluctuations will be determined by R_t and E_p. Transpiration rate was discussed previously. Components of R_t will now be examined to see how these compare between controlled environment and field conditions.

The resistance to flow in the soil can be approximated by (Gardner, 1960; Cowan, 1965; Campbell et al., 1976; Papendick and Campbell, 1974);

$$R_s = \frac{(\Psi_s^{1-n} - E_p B)^{1/1-n} - \Psi_s}{E_p} \tag{5}$$

where

$$B = \frac{(n-1) \ln (r_r^2 \pi L)}{4\pi D k_s Z \Psi_e^n} \tag{6}$$

and Z is rooting depth, D is root density (length of root per unit volume of soil), r_r is the mean radius of absorbing roots, Ψ_e is the air entry water potential (potential at which the soil just begins to desaturate) for the soil, k_s is the saturated hydraulic conductivity of the soil and n is an empirical constant from the hydraulic conductivity expression:

$$K = k_s (\Psi_e/\Psi)^n \tag{7}$$

Note that R_s depends on E_p (because hydraulic conductivity depends on potential gradient). Soil resistance also depends on pot dept and root density as well as the hydraulic properties of the rooting medium.

The root resistance in eq. 4 is the resistance to flow through the cortex and endodermis to the root xylem. The endodermis apparently is the most important barrier to water movement within the root because water flow is restricted to a symplasmic pathway in this tissue. For this reason the endodermal resistance is controlled in part by root respiration rates. Soil aeration and temperature are therefore important in determining the magnitude of this resistance. Soil temperature should be similar between field and controlled environment conditions and adequate aeration should be assured (pots well drained) for proper control of R_r.

The xylem and leaf resistances of healthy plants should be re- lated to leaf area index, (LAI) so if LAI is similar for field and controlled environment conditions, these resistances should be similar.

Herkelrath, Miller, and Gardner (1978) suggested that a contact resistance between the root and the soil should be included in calculation of R_r. Contact resistance has not been included in this analysis, though it should be included in a more complete analysis in the future as contact resistances are better quantified.

For lowered (more negative) soil water potential to result in plant water stress, either the soil water potential, Ψ_s, must become a significant fraction of the leaf osmotic potential, Ψ_π, or the soil resistance must become a significant fraction of the overall resistance to water flow in the soil-plant system. The fact that plant growth is reduced under field conditions at soil water potentials above -0.8 bar (tensiometer range) (Taylor and Ashcroft, 1972) indicates that the latter is likely the case in the field.

Fig. 2B shows R_s/R_t as a function of soil water potential for a silt loam soil with root densities and depths typical of field conditions. Figure 2A shows similar relationships for higher rooting density and shallower root depth typical of a pot experiment. Here the plant is able to extract water to potentials around -4 bars before the external resistances become significant compared with extraction to only -0.5 bar at the lower root density typical of field conditions. Thus, unlike the field situation, stress on plants grown in small pots is imposed first by lowering of soil water potential (rather than restricting flow) and finally by an abrupt increase in resistance after most of the soil water has been depleted. This gives the plant little time to adjust to the stress condition, so the usual responses to stress that one sees in the field (leaf abscission, osmotic adjustment, changes in morphology of new leaves) cannot occur.

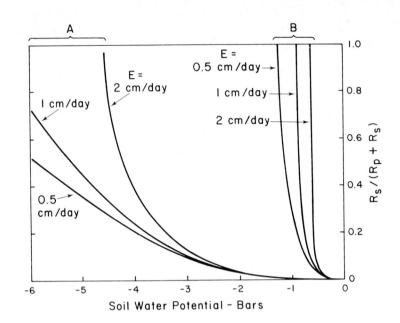

FIGURE 2. Ratio of soil resistance to soil plus plant resistances as a function of soil water potential for: (A) conditions typical of a controlled environment pot experiment (15 cm deep pot, root density = 10 cm^{-2}), and (B) deep rooted crop in the field after surface moisture is depleted (150 cm root depth, root density = 0.1 cm^{-2}). Soil hydraulic properties are as in Campbell et al. (1976).

MEASUREMENT OF PLANT WATER STATUS

Once an experiment has been designed using the guidelines just presented, the investigator should then assess the success or failure of the treatments that have been applied. If the objective of the experiment is to simulate field conditions, then field data are needed for comparisons. Measurements of diurnal and seasonal variations in leaf water potential, osmotic potential and stomatal resistance, soil water potential, soil temperature, and transpiration rate could be used. Equipment for making these measurements is readily available and generally not difficult to use.

Perhaps the most reliable and accurate method for routine measurement of leaf water potential and osmotic potential is the

pressure chamber (Scholander et al., 1965). To make a measurement, a leaf or portion of a leaf is excised from the plant and
inserted into a pressure chamber with a petiole, twig, midrib, or
other xylem containing part extending through a seal to the outside of the chamber. Pressure is applied to the leaf until free
water appears in the xylem, indicating that the xylem water
potential (Ψ_x) is zero. When $\Psi_x = 0$, $\Psi_\ell + P = 0$ so $\Psi_\ell = -P$, where
P is the pneumatic pressure applied to the leaf. For pressure
chamber measurements to be accurate, precautions must be taken to
assure that the leaf water content does not change during the measurement and that there is no physical damage to the leaf tissue
enclosed in the chamber. Gandar and Tanner (1976) recommend that
leaves be wrapped in a damp cloth before excision to minimize
evaporation. The cloth is left in place while a reading is taken.
Placing water in the pressure chamber does not completely stop
evaporation because temperatures in the chamber increase rapidly
with pressurization, causing a drop in humidity. Care must also
be taken to be sure that water is not lost from the xylem through
over-pressurizing the leaf. The rate of pressure increase must be
slow enough to maintain water potential equilibrium throughout the
leaf. This can be checked by increasing the pressure to an end
point, decreasing it until water disappears from the xylem, then
increasing it again to see if the same end point is obtained. If
the second reading is lower than the first, the rate of pressurization was too high.

Osmotic potential can be measured with the pressure chamber
using the method of Tyree and Hammel (1972). When turgor pressure
is zero, osmotic potential is equal to Ψ_ℓ, which is measured by
the pressure chamber. To use this method, leaf water content is
reduced until turgor is zero, and then measurements (of osmotic
potential) on the flaccid tissue are used to extrapolate back to
obtain the osmotic potential of the tissue at its initial water
content. Since the reciprocal of osmotic potential varies linearly

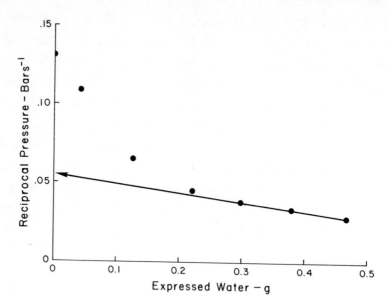

FIGURE 3. Water release curve for Prunus laurocerasus showing extrapolation to zero expressed water for osmotic potential determination. The osmotic potential computed using the extrapolation from the final three points is -18.1 ± 0.2 bars.

with water content, a plot of reciprocal chamber pressure for the flaccid tissue vs. water content is used for the extrapolation.

Operationally, the method is carried out as follows: After an initial pressure reading is made, some water is forced out of the leaf, either by over-pressurizing it in the chamber or allowing water to evaporate from the leaf. For large leaves, expressed water can be collected from the petiole or twig in a small vial or tube containing absorbent tissue, and the amount determined by weighing. For smaller leaves, the entire leaf must be weighed to determine changes in water content with sufficient accuracy (Campbell et. al., 1979). Measurements of leaf water potential and the mass of expressed water are continued until several measurements are obtained when the leaf is flaccid. Reciprocal pressure is then plotted as a function of mass of water lost from the tissue (Fig. 3). An extrapolation of the linear part of the curve back to zero expressed water gives the reciprocal of the osmotic pressure.

Water potential and osmotic potential can also be measured using thermocouple psychrometers. *In situ* leaf psychrometers (Neumann and Thurtell, 1972: Campbell and Campbell, 1974) allow non-destructive determination of leaf water potential, but require more care for accurate measurements than do pressure bomb measurements. Thermocouple psychrometer measurements on excised tissues are subject to error (Baughn and Tanner, 1976) and should be avoided whenever possible.

Thermocouple psychrometers work well for measuring osmotic potential. Leaf tissue is frozen (preferably with liquid nitrogen) to destroy cell membranes, and sap is squeezed onto a filter paper disk which is equilibrated and read in a psychrometer chamber. This method can be very rapid (1 sample per 2-3 min.) using a commercial (Wescor, Inc., Logan, Utah, Model C-52) sample chamber, but is subject to a potentially serious systematic error. Apoplastic water (water in cell walls) contains almost no solutes. This water mixes with cell solutes when the tissue is frozen and dilutes the cell sap (Boyer, 1972; Tyree, 1976). Apoplastic water fractions can be as high as 0.3 to 0.5 in some species (Hellkvist, Richards, and Jarvis, 1974); (Campbell *et al.*, 1979) so psychrometer measurements of osmotic potential can be substantially in error for these species. Psychrometer measurements can be corrected for this effect if an estimate of the apoplastic water fraction is available (Campbell *et al.*, 1979).

Soil water potential can be measured with a tensiometer in the 0 to -0.8 bar range and with *in situ* soil psychrometers (Rawlins and Dalton, 1967) in the range -1 to -50 bars.

Stomatal diffusion resistance can be measured in several ways. A thorough review of methods for making the measurement is available (Kanemasu, 1975). A simple mass flow porometer (Alvim, 1965) might be as useful as a diffusion porometer for comparing field and controlled environment plants where it is not necessary to know the actual diffusion resistance.

REPORTING MEASUREMENTS

The measurements that are most likely to indicate success or
failure in achieving good simulation of field conditions are
osmotic potential and diurnal variation in leaf water potential.
Certainly, pot dimensions and transpiration rates should be reported. Measurements that possibly are not essential, but would
be "nice to have" are leaf area index, root density, soil hydraulic and water retention characteristics, and soil temperature. In
any water stress study, soil water content or water potential
throughout the stress cycle should be reported. If water content
is reported, a moisture release curve for the potting medium
should also be published.

CONCLUSIONS

The examples given and the principles discussed here suggest
several reasons for the general failure of water relations of
plants in controlled environments to resemble water relations of
field grown plants. The most important reason may be failure to
consider and match the dynamics of water transport in the soil-
plant-atmosphere system in field and controlled environment conditions. Another reason may relate to light levels and photosynthesis rates that control osmotic potentials.

Researchers may continue to be reluctant to change methods because of the difficulty and expense involved in using larger pots,
better potting mixes, more measurements, and better control to
more readily simulate field conditions. These are certainly valid
concerns. On the other hand, if, because of the experimental design, an experiment has little chance of answering the questions
posed, the investigator has no choice but to modify the design.
Hopefully some of the principles and methods outlined here will be
useful in arriving at suitable experimental systems.

REFERENCES

Alvim, P. (1965). A new type of porometer for measuring stomatal
    opening and its use in irrigation studies. *In* "Methodology
    of Plant Eco-physiology " (F. E. Eckhardt, ed.) pp. 243-247.
    UNESCO, Paris.

Baughn, J. W., and Tanner, C. B. (1976). Excision effects on leaf
    water potential of five herbaceous species. *Crop Sci. 16*,
    184-190.

Boyer, J. S. (1968). Relationships of water potential to growth
    of leaves. *Plant Physiol. 43*, 1056-1062.

Boyer, J. S. (1972). Use of isopiestic technique in thermocouple
    psychrometry III. Application to plants. *In* "Psychrometry in
    Water Relations Research" (R.W. Brown and B.P. Van Haveren,
    eds.) pp. 220-223. Utah Agr. Exp. Sta. Logan, Utah.

Campbell, G. S. (1977). "An Introduction to Environmental Bio-
    physics." Springer-Verlag, New York.

Campbell, G. S., and Campbell, M. D. (1974). Evaluation of a
    thermocouple hygrometer for measuring leaf water potential *in
    situ*. *Agron. J. 66*, 24-27.

Campbell, G. S., Papendick, R. I., Rabie, E., and Shayo-Nhowi, A.
    J. (1979). A comparison of osmotic potential, elastic modu-
    lus, and apoplastic water in leaves of dryland winter wheat.
    Agron J. 71, 31-36.

Campbell, M. D., Campbell G. S., Kunkel, R., and Papendick, R. I.
    (1976). A model describing soil-plant-water relations for po-
    tatoes. *Amer. Potato J. 53*, 431-441.

Cowan, I. R. (1965). Transport of water in the soil-plant-atmos-
    phere system. *J. Appl. Ecol. 2*, 221-239.

Gandar, P. W., and Tanner, C. B. (1976). Potato leaf and tuber
    water potential measurements with a pressure chamber. *Amer.
    Potato J. 53*,1-14.

Gardner, W. R. (1960). Dynamic aspects of water availability to plants. *Soil Sci.* *89*,63-73.

Hellkvist, J., Richards, G.P., and Jarvis, P.G. (1974). Vertical gradients of water potential and tissue water relations in Sitka spruce trees measured with the pressure chamber. *J. Appl. Ecol.* *11*,637-667.

Herkelrath, W. N., Miller, E. E., and Gardner, W. R. (1977). Water uptake by plants: II. The root contact model. *Soil Sci. Soc. Amer. J.* *41*,1039-1043.

van den Honert, T. H. (1948). Water transport in plants as a catenary process. *Disc. Faraday Soc.* *3*,146-153.

Hsiao, T. C. (1973). Plant responses to water stress. *Annu. Rev. Plant Physiol.* *24*,519-570.

Kanemasu, E. T. (ed.). (1975). Measurement of stomatal aperture and diffusive resistance. Bull. 809, Washington Agric. Res. Center, Pullman, Washington.

Neumann, H. H., and Thurtell, G. W. (1972). A Peltier cooled thermocouple dew point hygrometer for *in situ* measurement of water potentials . *In* "Psychrometry in Water Relations Research" (R. W. Brown and B. P. Van Haveren, eds.), pp. 103-112. Utah Agr. Exp. Sta., Logan, Utah.

Papendick, R. I., and Campbell, G. S. (1974). Water potential in the rhizosphere and plant and methods of measurement and experimental control. *In* "Biology and Control of Soil-Borne Plant Pathogens" (G. W. Bruehl, ed.), pp. 39-49. Amer. Phytopath. Soc., St. Paul, Minnesota.

Papendick, R. I., and Cook, R. J. (1974). Plant water stress and development of Fusarium foot rot in wheat subjected to different cultural practices. *Phytopathology 64*,358-363.

Rawlins, S. L., and Dalton, F. N. (1967). Psychrometric measurement of soil water potential without precise temperature control. *Soil Sci. Soc. Amer. Proc. 31*,297-301.

Scholander, P. F., Hammel, H. T., Bradstreet, E. D. and Hemming-
    sen, E. A. (1965). Sap pressure in vascular plants, *Science*
    148,339-346.
Taylor, S. A., and Ashcroft, G. L. (1972). "Physical Edaphology."
    W. H. Freeman., San Francisco, Calif.
Tyree, M. T. (1976). Negative turgor pressure in plant cells:
    fact or fallacy? *Can. J. Bot.* *54*,2738-2746.
Tyree, M. T., and Hammel, H. T. (1972). The measurement of tur-
    gor pressure and the water relations of plants by the pres-
    sure bomb technique. *J. Exp. Bot.* *23*,267-282.

WATERING: DISCUSSION

THURTELL: Would measurements of stomatal resistance be useful for ensuring that plants grown under controlled environments are similar to plants grown in the field?

CAMPBELL: Yes, stomatal resistance measurements would be recommended strongly, for there often are significant differences between stomatal resistances of plants grown in controlled environments and those grown in the field. At night, stomatal resistance in controlled environments is usually less than in field grown plants, and during the day stomatal resistance is usually greater in controlled environments than in the field.

RAWLINS: I would encourage measurements of both plant water potential and stomatal resistance because we found, in long period growth studies with cotton, that both measurements are necessary to explain the variations in growth under different environmental levels. Plant water potential was similar in plants grown under different humidity conditions, but stomatal resistance was quite different as the stomata adjusted to the changed humidity levels.

SPOMER: Although it is difficult in chambers to duplicate the soil water stress conditions of the field, we should not ignore measuring the soil water stress. What kind of soil water measurements should be recommended in guidelines?

RAWLINS: The tensiometer is unexcelled for measurement at high soil moisture levels except that there may be problems in very coarse media. With low soil moisture contents psychrometers are useful. One should not disregard measurement of salinity in soils and conductivity meters can effectively monitor the salinity or osmotic potential of the soil.

CAMPBELL: For media moisture determinations, all that is really needed is a moisture release curve for the medium obtained by water content measurements. This would be simpler than taking tensiometer and psychrometer readings. The electrical conductivity measurements can be taken on an extract of saturated media.

FRANK: Is is possible to maintain containers at less than field capacity without going through drying cycles?

SPOMER: I have not found it possible to maintain a constant soil moisture level by the tension method, and thus have worked with a water content system. This involves using a mixture of particular proportions of fine and coarse particle materials in shallow containers. This system has been somewhat effective but there is some fluctuation in soil moisture content and the system also requires very frequent watering to minimize the fluctuations. I do not believe it is possible to maintain effectively a constant soil moisture tension.

GARDNER: Rawlins indicated in his presentation that the matrix forces were dominant in redistribution of water applied to containers. This will be the case with frequent and light waterings. However, with less frequent and heavy waterings, gravitational forces are dominant in the distribution of water. Thus with less frequent waterings it is more necessary to have uniform water applications at the surface if uniform distribution of water within the soil is to be achieved. Also with coarse textured media, as with gravel culture, gravitational forces are much more significant than matrix forces in water distribution.

SPOMER: I would also like to reemphasize that perched water tables of container grown plants in growth chambers and greenhouses significantly affect the water relations of the plants. Even with very coarse textured materials, if the containers are only a few centimeters deep, the media will be saturated with water when the water table is at the bottom of the container.

KRIZEK: Watering and nutrition of plants should be in relation to the time and length of the light period. Automatic watering systems provide much greater flexibility and precision than is
possible with personnel that do watering and are commonly geared
to an 8 AM to 4:30 PM day.

GLADON: Has anyone studied starch-based polymers that hold
many times their weight in water as a way of buffering the cycling
of water?

PARSONS: The starch-based polymers are still under development. We have had problems in obtaining uniform distribution of
the polymer in the soil mix. The polymer tends to clump and root
growth is not as good in those pockets. Polymers do tend to keep
plants turgid for a long period of time after a watering. The
cost of these is quite high, however.

SPOMER: The recorded water contents of certain organic materials, perlite, and calcined clay often contain significant
quantities of water that may not be available to plants.

PALLAS: We have found that peat-vermiculite mixtures provide
good media for research studies, although we have noted that
different lots of commercially prepared mixtures varied because
the manufacturer made changes in the quantity of added nutrients.

PALLAS: One finds that there are endogenous rhythms in transpiration, photosynthesis, and other plant processes, so that the
time of day for monitoring a plant process is quite critical. Also short period cycling in transpiration can occur. We have found
short period cycling in peanut plants when soil water stress exceeds one bar.

SALISBURY: I would also emphasize the existence of rhythmic
change in stomatal resistance, leaf orientation, pigment concentrations, nucleus size, and photosynthetic rates. We must be
aware of the role of the clock in the functioning of the plant
for it is an added complication in obtaining accurate measurements.

FRANK: Has stomatal cycling been observed in osmotically-adjusted plants, eg. plants not grown under ideal conditions?

KAUFMANN: Many people have felt that short period stomatal cycling is limited to plants grown in controlled environments. However, we noted that transpiration, photosynthesis, and even trunk diameter of trees of citrus maintained in the field, exhibit short period cycling. We don't really know how common this phenomenon is in the field.

COYNE: We recorded short period cycling of stomatal resistance in the field with ponderosa pine and snapbeans. Cycling has been observed when using leaf cuvettes on plants and maintaining an environment in the individual cuvettes that is different than the environment around the rest of the plant. The cycling appears to involve an overshoot mechanism that occurs as the plant adjusts to the modified environment of the cuvette. The cycling damps out eventually.

KRIZEK: When making stomatal resistance measurements, it is necessary that reporting include where, when, and how these are taken. There are very large differences between abaxial and adaxial surfaces, gradients on different leaves of a plant, time of day that measurements are taken, and variations associated with watering of the containers.

HELLMERS: Air movement rates in controlled environments do not duplicate those in the field and, therefore, water relations of plants in the field can not be effectively simulated. Particularly the necessity for significant air movement in chambers means that calm periods of field environments cannot be duplicated in controlled chambers.

PRECISION AND REPLICATION

C. H. M. van Bavel

Department of Soil and Crop Sciences
Texas A & M University
College Station, Texas

INTRODUCTION

It is certain that the diversity in background of individuals
who use controlled environments for plant research has resulted
in a confusion of ideas and concepts on required and attainable
levels of precision and on the role of replication. Hopefully,
this discussion will help to reduce the confusion, and lead to
adoption of rational and cost-effective attitudes.

First of all: what is precision? It is a measure of our
ability to reproducibly quantify a property or a process. For
this measure we can choose the relative standard error of the
mean of repeated observations. But, in the real world, properties
and processes vary with time, implying that, by merely repeating
observations, we cannot increase precision to a limit determined
by observational and instrumental error only.

Even though we may not ignore the latter form of error, one
purpose of environmental control is to increase precision by
relying on artificial rather than natural sources of light, heat,
CO_2, and water, and thereby to reduce that variation with time,
which we realize to be the main source of imprecision. In

323

addition, we use devices for corrective action when the difference
between successive measurements exceeds a certain amount. This
technology is discussed in other chapters, with reference to
specific environmental parameters.

Before further analyzing precision in environmental control,
it should be pointed out that obtaining a certain degree of
control precision is necessary, but it is not a sufficient condition for obtaining meaningful results. This additional requirement is *accuracy*.

Accuracy refers to a consistent agreement with reality;
therefore, accuracy in measurement makes it possible to combine
the data, results, and conclusions of experiments conducted in
various places at various times to arrive at generally valid
concepts. Accuracy may well be called the stepchild of biological research. We fail to ingrain students with the idea that the
degrees, millivolts, or CO_2 concentrations that are indicated on
instrument dials or shown by printers are, in fact, often
incorrect. Consider, for example, the common pneumatic air
thermostat. Many of these can provide full proportional action
over a range of 2.0°C, with a dead band of 0.2°C. Actually most
thermostats are inaccurate by two or three degrees C. The answer
to this problem is, of course, calibration: Comparison against a
standard under closely controlled conditions. This calibration
should preferably be more precise than that of the controlled
environment in which the calibrated instrument is to be used.
Further, it is advisable to duplicate calibration readings until
a point of diminishing returns is reached, as indicated by a
standard error of the mean that approaches the readability of the
instrument, for instance, 0.2°C on a thermometer, marked off in
degrees.

From the previous considerations it should be concluded that
precision need not be better than accuracy, as our ultimate goal
is general validity rather than specific repeatability. Calibration procedures are specific for each parameter or process.

For orientation purposes, a provisional estimate of generally
attainable accuracy, at normal levels of variables or functions,
is as follows:

air temperature	0.3°C
soil temperature	0.2°C
organ temperature	0.5°C
dewpoint temperature	0.2°C
relative humidity	0.02 or 2%
PAR	20 μ E m^{-2}s^{-1} (400-700 nm)
short-wave irradiance	20 W m^{-2} (400-2500 nm)
net radiant balance	50 W m^{-2} (all wavelengths)
air speed	0.1 m s^{-1}
CO_2 concentration	
absolute	0.5 ppmv
differential	0.2 ppmv
flow rate	2%
plant water potential	2 bars (200 kPa)
soil water potential	0.1-0.5 bar (10-50 kPa)
transpiration	0.02 mm hr^{-1}
CO_2 exchange rate	0.02 g m^{-2}hr^{-1}

In the last two cases, the gas exchange rates can be understood
as applied to a unit leaf area, or a unit "ground" area, when
experiments are conducted with closed stands, as found under
standard cultural management.

PRECISION OF ENVIRONMENTAL CONTROL

The physical plant environment is generally characterized by
the temperature, humidity, CO_2 content, and velocity of the air,
by the incident radiant energy which may impinge on plant parts
from any direction, and by temperature, water potential, pH,
salinity, composition, and aeration of the soil.

It is common practice to measure any of these parameters, and
usually all of them, at a single point in the experimental space.
A first recommendation is to study, with the experimental plants
in place, the *spatial variability*. Such an effort will often be
informative in showing that spatial variability considerably
exceeds both the precision and accuracy of measurement.

Specific problems of lack of uniformity in the various
radiant fluxes have been discussed by others in this conference.
The often observed lack of uniformity of air properties - humidity,
CO_2 content, temperature, and air speed - are due to the common
practice of providing only minimal air speeds. It was argued
earlier (van Bavel, 1973) that, in order to obtain normal mass
exchange coefficients between the leaves and air, the air speed
in an artificial environment should be at least of the order of
1 m s^{-1}. It is clear that a ventilation rate of that magnitude
will also greatly improve uniformity of the environment, in
addition to being realistic.

Nevertheless, temperature differences of 1°C between incoming
and outgoing air, are commonly found when a normal radiative load
is imparted to the plants in the environment, and when a realistic
rate of air movement is provided. Likewise, differences of 2°C
may typically be found between leaf temperatures, depending on
individual leaf location, attitude, and/or shading. Comparable
variation is to be expected for air humidity and for transfer
coefficients.

Our experience is that the unavoidable lack of uniformity within
the experimental space, and between test objects, exceeds the
requirements often specified by plant scientists and exceeds the
specifications indicated for chambers as measured at a single
point.

PRECISION OF BEHAVIORAL CONTROL

Although we concentrate the engineering effort on environmental control, it should be stressed that our purpose is to control plant behavior. If we cannot elicit the same response from a given plant species for a given stage of development and previous history, by repeated or continuous exposure to the same set of environmental conditions, we will not be able to define the response function, when one or more environmental parameters are varied.

Plant behavior can be thought of as the result of five primary processes: photosynthesis, respiration, transpiration, mineral uptake, and development. The first three are strongly and directly affected by radiant energy flux, ambient CO_2 level, and humidity. These three environmental factors also affect stomatal opening; hence they indirectly influence photosynthesis and transpiration. We also know that respiration depends to a large measure upon photosynthesis, as it reflects the rate at which photosynthate is used in further biosynthesis.

The traditional approach to environmental control has emphasized light quality and photoperiod, temperature, and mineral substrate balance, at the expense, or even neglect, of light uniformity throughout the test space, adequacy of radiant energy levels, CO_2 concentration, and humidity. The precision of behavioral control under such conditions may have been adequate with respect to development and mineral uptake, but not to the three processes that regulate the carbon and water economy of plants.

But a question should be asked about how and to what precision we can know that plant behavior is being controlled on a real-time basis. Transpiration can sometimes be monitored by periodic or continuous weighing of entire plants or small plant stands. With available techniques, an hour and a relative error of 10%

are the approximate limits of resolution. Indirectly, we may
gauge the rate of transpiration by monitoring leaf temperatures,
epidermal resistance, or leaf density by beta gauging. All of
these indirect techniques are tedious.

A more elegant approach is to measure the water balance of the
entire system. In once-through systems the humidity differential
and the flow rate can be recorded for calculation of water loss,
but the precision is no more than about 10%, and the time constant
is of the order of one hour. In entire systems there is no way
to differentiate transpiration rates of individual plants and
therefore the average behavior measured may obscure great
differences in rates between individuals. Also, an arithmetic
average could mask non-linear response at the organ or plant level
over a period of time. Therefore, regardless of tedium or cost,
exploratory continuous measurements of transpiration rates of
individual leaves or plants are necessary, so that we may know
whether and to what precision we are controlling the process, by,
for example, controlling air humidity. Similar comments can be
made with regard to the other two gas exchange processes: photosynthesis and respiration.

It can be demonstrated from simple models of canopy behavior
that Michaelis-Menten type equations decribe many environmental
effects: air movement vs. gas exchange rate, saturation deficit
vs. transpiration, light level and CO_2 level vs. photosynthesis
and the ensuing respiration. Further it appears that only rarely
can one produce "saturating conditions"; that is, a level of the
environmental parameter at which its precise value is not critical.
Rather, one operates generally on a part of the response curve
where the slope is appreciable.

Only with regard to light quality, mineral nutrition, and soil
water and soil air supply can the assumption be made, and then
only with caution, that a "non-limiting" situation prevails.

Environmental temperature is a factor by itself - it always plays a significant role in the functioning of organs and in many life processes. Thus, the historical emphasis on its control was and remains justified, although the necessary distinction has not always been made between air temperature and that of aerial plant organs and tissues.

REPLICATION

We have seen that, with moderate demands on precision and accuracy of environmental control - as long as it is comprehensive - a given artificial environment can be duplicated at any time to a closer degree than is needed in view of other causes of divergent plant behavior. Therefore, and from this viewpoint alone, experiments need not be repeated, in contrast with tests performed out of doors, or with partial inclusion of outdoor effects, as in greenhouses.

On the other hand, there generally is no assurance that the nature or condition of test plants can be duplicated to a similar degree. This consideration requires replication in time. The only difference with the parallel situation in plant research in the field is that the time can be chosen at will.

Secondly, there is ample reason to expect, from physical and biological divergence within a controlled environment, that individual plants or groups of plants will function differently, depending on what and where they are. Therefore, the test objects must be replicated so the investigator will know the magnitude of the error so induced and the limitations it imposes on precision of the experimental outcome. Again, the field agronomist or forester faces this same problem, in that individual plots or parcels are different with regard to soil conditions or microclimate.

An extensive literature and forty years of experience exist
on how to design experiments with plants in the field, how to
estimate the contributions of different sources of error, and how
to interpret the results. I believe that these statistical
techniques are equally pertinent to studies in controlled environments, even though one source of environmental variation is or
at least can be effectively removed. These techniques allow the
investigator to decide on the optimum number of replications and,
in general, to design experiments efficiently. This is not new
to crop physiologists, but it may be appropriate to make this
suggestion to all plant scientists who wish to make the best use
of controlled environments.

It is inappropriate to give rules of thumb or specific
suggestions for replication other than to indicate that, no
matter how precisely and accurately the environment is controlled,
there will be need for replication. This consideration implies
that there are rational limits to the required degree of control
in controlled environments.

REFERENCES

van Bavel, C. H. M. (1973). Towards realistic simulation of the
    natural plant climate. *In* "Plant Response to Climatic
    Factors" (R. D. Slatyer, ed.), pp. 441-446. Proc. Uppsala
    Symp. (1970) UNESCO.
van Bavel, C. H. M. (1975). Design and use of phytotrons in
    crop productivity studies. *Phytotronic Newsletter 10,* 16-22.

PRECISION AND REPLICATION: CRITIQUE I

Henry J. Kostkowski

National Bureau of Standards
Washington, D. C.

This paper consists of: (1) a summary of the standards and
calibrations pertinent to measurements of controlled environments
and available from the National Bureau of Standards, (2) estimates
of the state-of-the-art accuracy available with commercial instruments, (3) a discussion of why an accurate calibration does not
insure an accurate measurement, and (4) brief remarks about the
uncertainty of error estimates.

The National Bureau of Standards (NBS) is the primary
government laboratory concerned with physical standards and
measurements.   It is responsible for realizing, maintaining, and
disseminating the basic physical standards required in the United
States for science and technology.

Tables 1 to 5 contain information on the standards and
calibrations currently offered by NBS that are pertinent to
measurements of controlled environments.   Also included is an
estimate of the current state-of-the-art accuracy for measurements
with calibrated, commercially available instruments and a list
of useful publications.   This material was obtained from the NBS
divisions listed in the tables.   Investigators are invited to
contact these offices for any additional information they might
need relative to NBS services.

A number of the entries in the tables require an explanation
or some additional comments.

TABLE 1. *NBS Standards, Calibrations and Other Information of Interest for Measurement of Optical Radiation in Controlled Environments*

STANDARDS

Quantity	Description	Uncertainty	Cost	NBS Contract
Spectral irradiance ($\mu W\ cm^{-2}\ nm^{-1}$)	8A - 120V Tungsten lamp, 250-1600 nm	2.6% - 1.2% respectively	$1655	Radiometric Physics Division (Attn: D. A. McSparron)
	300 mA - 120V D_2 lamp, 200-350 nm	6%	$790	
Spectral responsivity ($A\ W^{-1}$ and $A\ cm^2\ W^{-1}$)	Silicon photodiode and amplifier 380-880 nm	1.5%	$500 (rental)	Radiometric Physics Division (Attn: E. F. Zalewski)
	257-364 nm and a single point at 1064 nm	5%		

STATE-OF-THE-ART UNCERTAINTY WITH CALIBRATED COMMERCIAL INSTRUMENT

2 - 25% depending on character of the radiation and measurement conditions.

REFERENCES

Belanger, B. C., ed. (1978). Calibration and related measurement services of the NBS. NBS Publ. 250.

EOSD (1978). Electro-optical systems design. Vendor Selection Issue. 222 West Adams, Chicago, Illinois.

Kostkowski, H. (1977). The national measurement system for radiometry and photometry. NBS Interagency Rept. 75-939.

Nicodemus, F. E., ed. (1976-1978). Self-study manual on optical radiation measurements. NBS Tech. Notes 910-1, -2, and 3.

The spectral responsivity standard referred to in Table 1 involves the rental, for two weeks, of a silicon-photodiode, broad-band radiometer calibrated for spectral responsivity at the wavelengths indicated. Renting the unit obligates the individual to participation in an intercomparison program including returning certain data to NBS.

In recent years there has been a large increase in all types of commercial radiometers including those with pyroelectric, silicon, or electrically calibrated detectors; with portable single and double monochromators; and with convenient-to-use electronic systems or even microprocessors for automatic control and data processing. These instruments significantly reduce much of the tediousness of radiometry, and many of them have a day-to-day reproducibility of about 1%. As shown in Table 1, even though radiometric standards with uncertainties of 1.2% to 2.6% are available for the ultraviolet (UV), visible and near infrared spectral regions, the state-of-the-art uncertainty when using these calibrated instruments may be and usually is much worse. This is due to the multidimensionality of radiometric measurements. For example, the character of the optical radiation in different growth chambers may vary considerably from each other and from that of a standard. In general, optical radiation varies with

TABLE 1 (continued)

REFERENCES[a] (continued)

Lind, M. A., Zalewski, E. F., and Fowler, J. B. (1977). The NBS detector response transfer and intercomparison package: The instrumentation. NBS Tech. Note 950.

[a]All NBS publications except NBS Interagency Reports are obtainable from the Supt. of Documents, U.S. Govt. Printing Office, Washington, D. C. 20402. The NBS Interagency Reports are available from the National Bureau of Standards, Washington, D. C. 20234.

position and direction (e.g. at the receiving aperture of a
radiometer) and with time, polarization, and wavelength; and the
responsivity of a radiometer often varies with these same parameters. These functional differences often degrade the measurement
accuracy significantly. Although 2% uncertainty is possible under
near ideal conditions, 25% is probably the best that has been
realized under the most unfavorable conditions (e.g. in measuring
solar terrestrial spectral irradiance at 290 nm). Under more
typical conditions, the state-of-the-art uncertainty is about 5%
in the visible and 10% in the UV. However, optical radiation
measurements in controlled environments are usually not performed
in a state-of-the-art manner so that poorer results are expected.

NBS' basic calibrations for temperature do not include a
thermometer that is particularly convenient for use in measurements of controlled environments. As indicated in Table 2, a
bead-in-glass thermistor should be considered for the measurement
of air and soil temperatures. These devices have time constants
of a few seconds or less and are capable of yielding measurements
to the accuracy with which they are calibrated, at least to a
few millidegrees. Experimenters will have to calibrate the
thermistors themselves, but this should be rather straightforward
if an NBS calibrated thermometer and a stirred water bath are
used.

NBS recently developed a new humidity calibration facility
capable of providing calibrations in relative humidity from 3%
to 98% at temperatures of 0 to 80°C with an estimated uncertainty
of 0.2%. However the best commercially available hygrometers can
only be calibrated with an uncertainty of 0.5 - 1.5% in relative
humidity, due to limitations in the instruments themselves (Table
3).

The familiar sling psychrometer can be used with an uncertainty
of about 3% when the thermometers are accurately calibrated
(typically an error of about 0.15°C in wet-bulb temperature

TABLE 2. NBS Standards, Calibrations and Other Information
of Interest for Measurement of Temperature in Controlled
Environments

CALIBRATIONS

Instrument	Description	Uncertainty	Cost	NBS Contact
Mercury-in-glass thermometer	Total immersion type with graduations of 0.1°C or less	0.03 °C[a]	$28 per temperature point	Temperature Measurements and Standards Division (Attn: Jacquelyn A. Wise)
Thermocouple	Type T (Copper-Constantan)	0.1°C[a]	$68 per temperature point	

STATE-OF-THE-ART UNCERTAINTY WITH CALIBRATED COMMERCIAL INSTRUMENT

Convenient thermometers for air and soil temperatures are bead-in-glass thermistors (Temperature Measurements and Standards Division, Attn: B. W. Mangum) which are primarily limited by the uncertainty of the above calibrations for mercury-in-glass thermometers or thermocouples.

REFERENCES

Belanger, B. C. (1978). Calibration and related measurement
    services of the NBS. NBS Publ. 250.

Roeser, W. F., and Lonberger, S. T. (1958). Methods of testing
    thermocouples and thermocouple material. NBS Circular 590.

Schooley, V. F. (1976). The national measurement system for
    temperature. NBS Interagency Rept. 75-932.

Wise, J. A. (1976). Liquid-in-glass thermometry. NBS Monograph
    150.

Wood, S. D., Mangum, B. W., Filliben, J. J., and Tillett, S. B.
    (1978). An investigation of the stability of thermistors.
    J. Res. Nat. Bur. Stand. 83, 247.

[a] The error associated with the stirred-liquid-bath facility
in which these calibrations are performed is 0.01°C.

TABLE 3. NBS Standards, Calibrations and Other Information
of Interest for Measurement of Humidity in Controlled Environments

CALIBRATIONS

Instrument	Description	Uncertainty	Cost	NBS Contact
Hygrometer	See references below	0.5 - 1.5%[a] (relative humidity)	Typically $400 for one ambient temperature	Thermal Processes Division (Attn: Saburo Hasegawa)

STATE-OF-THE-ART UNCERTAINTY WITH CALIBRATED COMMERCIAL INSTRUMENT

About 1% using high quality instruments such as aspirated psychrometers or dew point or electric hygrometers.

REFERENCES

ASTM (1978). Standard method of test for relative humidity by
    wet and dry bulb psychrometer E337-62. Annual Book of ASTM
    Standards, ASTM, 1916 Race, Philadelphia, Pennsylvania 19103

Hasegawa, S., and Little, J. W. (1977). The NBS two-pressure
    humidity generator, Mark 2. J. Res. Nat. Bur. Stand.
    (Physics and Chemistry)81A, 91.

Wexler, A. (1970). Measurement of humidity in the free atmosphere
    near the surface of the earth. Meteor. Monogr. 11, 262.

Wexler, A. (1979). A study of the national humidity and moisture
    measurement system. NBS Interagency Rept. 75-933.

[a]Error associated with the calibration facility is 0.2%.

results in an error in relative humidity of 1%); but with poor
technique, this can degrade to 4 or 5% (see ASTM Standard Method
E337-62 referenced in Table 3).

NBS currently has no appropriate standards for measurement
of atmospheric CO_2. However, plans exist for developing such
standards with an uncertainty of 0.1% (.3 ppmv) in about two
years (Table 4). Standards with a 1% uncertainty could be
developed much sooner. If these are adequate for controlled
environment measurements and a widespread need exists, it should
be brought to the attention of the Office of Standard Reference
Materials at NBS (George Uriano, Chief). Incidentally, NBS
is expecting to have a 1% CO_2-in-nitrogen standard, for CO_2
measurements on automotive exhausts, by the fall of 1979. However,
using these standards for determining CO_2 concentrations in air
with the frequently used non-dispersive infrared CO_2 detector
could result in an uncertainty of 2 or 3% in the measurements.
The reason for this is that the width of the CO_2 absorption lines
is different for CO_2 in air than for CO_2 in nitrogen, and this
difference could produce an additional uncertainty of 1 - 2%.
Using proper standards and great care, the non-dispersive infrared
instrument is capable of 0.1% uncertainty and a 0.5% uncertainty
is relatively easy to achieve.

As shown in Table 5, anemometer calibrations are available
from NBS, and the state-of-the-art uncertainty of air velocity
measurements with commercial instruments is about 1%.

Calibrations for water quantity or flow are not included in
the attached tables, but calibrations with an uncertainty of 0.13%
are available at NBS (Fluid Engineering Division, Attn: Kenneth
R. Benson). A 1% uncertainty can be realized without a calibration when using a good quality meter. However, because of changes
in time due to galvanic effects, the meter should be checked periodically using a second, infrequently used meter.

The third topic in this paper is the question of why an accurate calibration does not insure an accurate measurement. The

TABLE 4. NBS Standards, Calibrations and Other Information
of Interest for Measurements of CO_2 Concentration in Controlled
Environments

STANDARDS OR CALIBRATIONS   CONTACT

None currently available   Gas and Particulate
(see text for future plans)   Science Division (Attn:
                                      Ernest E. Hughes or
                                      William D. Dorko)

STATE-OF-THE-ART UNCERTAINTY WITH CALIBRATED COMMERCIAL INSTRUMENT

A non-dispersive infrared CO_2 detector will yield a determination
to about .1% (.3ppmv) if an appropriate standard is available.
CO_2-in-air mixtures currently available from commercial gas
producers have an uncertainty of 5 - 10%.

REFERENCES

Fastie, F. G., and Pfund, A. H. (1947). Selective infra-red
    gas analyzers. J. Optical Soc. Amer. 37, 762.

Pales, J. C. (1965). The concentration of atmospheric carbon
    dioxide in Hawaii. J. Geophys. Res. 70, 6053.

TABLE 5. NBS Standards, Calibrations and Other Information
of Interest for Measurement of Air Velocity in Controlled
Environments

CALIBRATIONS

Instrument	Description	Uncertainty	Cost	NBS Contact
Anenometer	$0.1\ ms^{-1}$ to $9\ ms^{-1}$	1%	$600	Fluid Engineering Division (Attn: Norman E. Mease)

STATE-OF-THE-ART UNCERTAINTY WITH CALIBRATED COMMERCIAL INSTRUMENT

High-quality anemometers of the Biram (vane) and hot wire types
will yield 1% measurements when appropriately calibrated and
approximately 10% without a calibration.

answer to this question is that the standard is usually somewhat different from the quantity being measured and the measuring instrument is sensitive to this difference. Using a CO_2-in-nitrogen standard rather than a CO_2-in-air standard when calibrating the non-dispersive infrared CO_2 detector was an example of this. A difference almost always exists between the standard lamp and the source lamps in controlled environments and is the reason, referred to earlier, why state-of-the-art measurements usually have a much greater uncertainty than either the uncertainty of the standard or the imprecision of the instrument. Two differences that are often the source of large errors in radiometry are differences in the spectral and directional distribution of the radiation.

Relative to differences in spectral distribution, one case was brought to the author's attention a few years ago by researchers working on the phototherapy of jaundice where a radiometer made an error of 10,000% even though calibrated with an accurate tungsten standard. The radiometer was being used to determine the irradiance between about 400 and 500 nm of a bank of fluorescent lamps. The instrument contained a blue filter that had a small near-infrared transmittance and a silicon detector that responded more strongly to the infrared than to the blue radiation. When a tungsten standard was used, where the ratio of red to blue radiation was very much greater than that from the fluorescent lamp, a tremendous measurement error resulted.

Differences in the directional distribution of the standard and the unknown radiation can also produce large errors. To

REFERENCES

Haight, W. C., Klebanoff, P. S., Ruegy, F. W., and Kulin, G. (1976). The national measurement system for fluid flow. NBS Interagency Rept. 75-930.

avoid these errors the radiometer's responsivity must not vary
significantly with the direction of the incident flux. This is
equivalent to saying the instrument must have a so-called cosine
correction. Since most radiometric standards fill a rather small
field of view, when a non-cosine-corrected instrument calibrated
with such a standard is used to measure radiant flux comprising
a large field of view, a large systematic error results.

The only situation in which an accurate calibration *insures*
an accurate measurement is when the radiation field of the standard and that being measured are identical. Otherwise, auxiliary
experiments must be performed to determine the effect of the
differences of the two fields on the measuring instrument. A
systematic way of addressing such problems in radiometry is
given in Chapter 5 of the NBS Self-Study Manual on Optical
Radiation Measurements referred to in Table 1.

The final topic to be covered is the uncertainty of error
estimates. This subject is rarely addressed; yet when comparing
results obtained by different experimenters it is generally
assumed that the error or uncertainty estimates made by the
experimenter are quite accurate. Nevertheless it has been the
author's experience that it is not unusual for error estimates
in the best radiometry performed to be incorrect by a factor of
two or more. This is really not surprising since the only
objective way of estimating systematic errors is to perform a
comprehensive set of experiments on the effects of varying all
the radiation, instrument, and environmental parameters. A
quantitative estimate of the systematic errors can then be made
in terms of the statistics of these experiments. However, such
comprehensive investigations are so difficult and time consuming
that they are rarely performed. Typically, estimates of individual systematic errors are made from "educated" guesses based on
one or two experiments and a few calculations. Then the individual errors are combined in quadrature (square root of the sum

of the squares) to obtain a total uncertainty for the reported measurement. There is no way of objectively estimating the accuracy of this total uncertainty, but it is not surprising when a new state-of-the-art measurement reveals an error *in this uncertainty* of 100% or more.

PRECISION AND REPLICATION: CRITIQUE II

P. Allen Hammer

Department of Horticulture
Purdue University
West Lafayette, Indiana

N. Scott Urquhart

Department of Experimental Statistics
New Mexico State University
Las Cruces, New Mexico

As van Bavel's presentation emphasized, demands by plant
scientists for control precision at a single point often far
exceed the control precision over the experimental space. However,
we should ask how this single point precision affects precision
over the experimental space before we assume these demands should
be relaxed.

VARIATION

Some expansion of discussion of replication is in order, but
we will discuss it in the larger context of experimental design.
Experimental design (blocking, replication, and randomization) is
as important in growth chamber studies as in greenhouse or field
studies. The growth chamber provides a known, researcher-
determined environment for plant studies. However, this does not
mean that sources of "unwanted" variation have been eliminated.
Appropriate experimental designs will organize the effect of

unwanted variation so that all treatments will be equally affected
by the unwanted variation; consequently such variation will have
no effect on comparisons between treatments.

Lack of uniformity among plants can be a consequence of this
unwanted variability. This has been determined among plants
within a chamber (Carlson, Motter, and Sprague, 1964; Collip and
Acock, 1967; Hammer and Langhans, 1972; Measures, Weinberger and
Baer, 1973; Lee, 1977; Rawlings, 1979), among plants grown in different chambers (Collip and Acock, 1967; Hammer et. al., 1978; Lee,
1977; Rawlings, 1979), and among plants grown in the same chamber
at different times (Hammer et. al., 1978; Lee, 1977; Rawlings,
1979). Differences in environmental conditions have been measured
within and between chambers (Kalbfleisch, 1963; Carlson, Motter,
and Sprague, 1964; Gentner, 1967; Hammer and Langhans, 1972; Measures, Weinberger, and Baer, 1973); Knievel, 1973; Tibbitts et.
al., 1976) and are probably responsible for much of the unwanted
variability and lack of reproducibility in plant growth within and
between chambers. Also, vibrations and handling of plants (Mitchell et. al., 1975) and contaminants within buildings and chambers
(Tibbitts et. al., 1977) may contribute unwanted sources of variation.

It is not possible in the short time available to cover all
the details concerning design of growth chamber studies. However,
several important points should be made. Van Bavel stated: "We
have seen that, with moderate demands on precision and accuracy
of environmental control - as long as it is comprehensive - a
given artificial environment can be duplicated at any time to a
closer degree than is needed in view of other causes of divergent
plant behavior." This statement is questionable, and there
certainly are enough data to suggest that time is indeed an important variable in growth chamber studies (Lee, 1977; Hammer et al.,
1978; Rawlings, 1979). In fact, it is so important that Lee
(1977) and Rawlings (1970) suggested, from very comprehensive
studies of uniformity at the North Carolina State University

Phytotron, that the between trials (or runs over time) variations[1] was more important (larger) than the between chamber variation.[1] They suggested blocking over trials (time) to account for this source of "unwanted" variability. When we use the term "block over time", we should be careful to clearly understand time as a block because time can index several things. Time can be chronological time (e.g. time of year, number of days), physiological time (maturity of a leaf, flowering), or time-related environmental variation. It is the time-related environmental variation that we should block against in growth chamber studies.

In order to more fully explain what has just been stated, and to offer some constructive suggestions for experimental design, the principles of design should be discussed first.

PRINCIPLES OF DESIGN

Fisher (1960) advanced three basic principles of experimental design: randomization, blocking, and replication. Since this initial work many books have been devoted to this topic [e.g. Cochran and Cox (1957), Federer (1955), and Kempthorne (1952)]. Consequently our comments can only highlight the principles; the next section interprets them for specific growth chamber experimentation.

Variation

All biological material exhibits variation (such as genotypic variability), even when plant material of only one cultivar is grown under apparently the same conditions. The previous section shows, however, that environments in growth chambers are not uniform throughout, regardless of the efforts made to control

---

[1]*Usually referred to as component of variance.*

them. This lack of uniformity in environments adds to the
genotypic variability. The design principles of *blocking* and
randomization provide tools for minimizing the impact of biological (genotypic) and environmental variation on treatment comparisons of interest.

To clearly see the need for randomization, suppose we have a
situation with biological but no environmental variation, and
have no way of predicting the biological variation before the
study begins. Suppose we unintentionally applied one treatment
to all of the smaller plants and another treatment to the larger
plants, and planned to record the effect of treatments on plant
growth. Our comparison of treatments would be confounded with,
i.e. mixed up with, initial plant size. In other words, if we
see a different response between the plants in the first treatment
and in the second treatment, we do not know whether to ascribe
this difference to the treatments or to the initial size of the
plants in each treatment. Randomization virtually eliminates
this problem. Randomization assures that each plant has the same
chance of receiving either of the treatments. Consequently, the
large and small plants have the same chance of influencing each
of the treatment means. In fact, randomization usually will
assign plants to treatments better than the investigator can
because he can perceive only a limited number of characteristics
of the plant while randomization assigns plants to treatments
without regard to any specific characteristics. It should be
remembered though that these comments apply when the investigator
has no specific information about the *organization of the
biological variation;* if we have such information we should use
blocking, an idea discussed below.

Randomization

Randomization has an important statistical consequence: it
is necessary for a valid estimate of the variation (variance)
among similar plants. This quantity is critical for making

comparisons between treatment means. Rarely will two treatments
have exactly the same means; when they differ, the estimate of
variance is central to evaluating how much difference between
treatment means might occur simply as a consequence of biological
variation. What could happen if the researcher assigned treatments
to plants rather than randomized them? To answer this, suppose he
balanced the assignment of plants to treatments so that the
average size of plants in each treatment was essentially equal
before he applied the two treatments. This will have two effects
from which randomization would protect him: (1) his estimate
of variance will be inflated because he will get pretreatment means
similar by increasing the difference between individual plants
within each of the treatments [if this does not make sense, take
some numbers and try doing what is suggested, or see Federer
(1955, p. 14)], (2) by removing all pretreatment difference in
plant size he will reduce the post treatment difference. Consequently, such nonrandom assignment of treatments to plants
reduces sensitivity two ways, by reducing the observed treatment
difference and by increasing the quantity used to measure the
biological variation. This illustration shows that nonrandom
assignment of plants to treatments can materially reduce a
researcher's ability to detect real differences; other types of
nonrandom assignment also can lead to biased results. THUS
RANDOMIZE.

How should an investigator actually accomplish the randomization? Suppose he has 20 plants numbered 1 through 20 to assign
to two treatments. He needs to use some random device or chance
mechanism to select 10 of these for the first treatment and
with the rest going to the second treatment. For example,
plastic discs numbered 1-20 can be put into a container, mixed
thoroughly, and then ten discs selected to give the plant
numbers for the first treatment. (Slips of paper tend to stick
together so they do not work as well as plastic discs.) Or an

ordinary deck of (new) playing cards marked 1-52 may be used.
The cards should be shuffled well, and cards taken from the top
of the shuffled deck until ten of the numbers 1-20 turn up.
These plants go to the first treatment and the rest go to the
other treatment. Or a table of random numbers, found in almost
any statistics textbook, can be used. This is done by entering
the random number table at some haphazardly chosen point and
using the numbers found there to go to another part of the table.
The investigator should find consecutive (non-overlapping) pairs
of digits in the tables; he should ignore those over 20, but
record the first ten up to and including 20. These are the
plants assigned to the first treatment. In other contexts a
researcher may need to repeatedly randomize a few things; for
two things, he should flip a coin and for 3, 4, 5 or 6 things,
roll a die - one of a pair of dice - and ignore the higher
numbers of dots for randomization of fewer than six things.
A pair of dice can be used only with great care, because some
numbers of dots (total of the two dice) occur with much higher
frequency than others.

Randomization and haphazard or unplanned assignment are not
equivalent. For example, an investigator could put the 20
plants of the previous paragraph in a row and select the first 10
for the first treatment. This would be a haphazard assignment.
In a particular case haphazard assignment may work as well as
randomization, but there is no assurance of how good it is. In
fact, randomization is much like insurance, you may not need it,
but if you need it, you need it badly! Thus randomize everything
you can.

These recommendations for randomization assumed no known
environmental variation, or recognized plant variation of a
known sort. No matter how diligently we try, environments still
vary somewhat, a point forcefully made in the first section of
these comments. BLOCKING provides the tool for dealing with

known sources of environmental and/or plant variation. A block
is a set of plants and/or microenvironments which is homogeneous
or internally similar. At least there is more similarity of
units in a block than between units in different blocks. Typically
a block of environmentally similar units will be contiguous and
close in space. In fact it often will consist of a square area.
Although square or nearly square blocks are conventional and
typical, the essential feature is that the environment be
relatively constant within the block. A block could be irregularly shaped if enough were known about the experimental area
to justifiably say that the irregularly shaped area contains a
more homogeneous environment than a square or rectangular area.
For example, extensive contour terracing has been done in the
dryland wheat country of Kansas and Nebraska. A study of cultural practices or cultivars of wheat justifiably could have
blocks laid out along the terraces. Such blocks would curve as
they followed a contour across a field. Similar blocking might
be appropriate in a growth chamber if contours of constant
performance had been established therein by a relevant uniformity
trial. (Note: the blocks defined here often are called repli-
cates in the agronomic literature. The usage here is consistent
with most of the statistical literature where a replicate concerns
an individual trial or plot.)

Treatments are assigned to locations in blocks so that each
treatment appears in each block exactly the same number of times,
usually once. The mechanics of the randomization would be accomplished as outlined previously, except that there would be a
separate randomization of treatments to locations in each block.
If additional similar size plants of the same cultivar are placed
in a growth chamber, their location in the chamber also should
be randomized. If on the other hand plants are moved into a
growth chamber that exhibit variation of consequence, say in
size, they should be grouped by size with a constant sized group

making up a block. In this case blocks would minimize the impact of both microenvironment and plant variation on the comparisons of treatments. The experimental design described here is called a RANDOMIZED COMPLETE BLOCK. It is complete in the sense that every treatment occurs in every block. Incomplete designs, namely ones where not all treatments appear in each block, do exist, but they usually require a fairly large number of blocks and must be set up and executed with care in controlled environments. In fact a researcher probably should seek a statistician's counsel before deciding on the use of an incomplete block design.

In the illustration above, each block consisted of constant sized plants in a homogeneous environment. In other words, two sources of heterogeneity were held constant in each block. The effect of the two sources of heterogeneity are confounded (inseparable). Confounding of sources of heterogeneity in construction of blocks is an acceptable experimental design practice, in contrast to the undesirable confounding of treatments with sources of heterogeneity. Sometimes unwanted sources of heterogeneity cannot be confounded effectively in the blocks. For example, assume that the variation among chambers was one source of heterogeneity and variation over several time periods was another; both of these sources of variation need to be dealt with in the design of the experiment. Designs for two-day elimination of heterogeneity do exist. The simplest and most common ones are called the LATIN SQUARE DESIGN and a slightly generalized version sometimes is called a LATIN RECTANGLE DESIGN. Both of these are illustrated in the next section. For a study of three air temperatures in three chambers over three time periods, the latin square design would require that each air temperature appear once in each chamber, and once in each time period. Such a design is restricted to having the same number of air temperatures, chambers, and time periods. The latin rectangle offers more flexibility by requiring that each air temperature appear the same number of times in each chamber, and

the same number of times in each time period. For example, a
2 x 4 latin rectangle could be used to compare two air temperatures in two chambers during four time periods.

Replication

The last major design component, REPLICATION, concerns
observation of several experimental units under the same treatment and environment. Here an experimental unit is the amount
of material to which a treatment was randomly allocated; typically
it is a plant, pot, or group of plants. The difference between
replicate observations should reflect individual plant variation
and microenvironmental variation not otherwise blocked out.

Replication has three major impacts: (1) it is required to
obtain a valid estimate of residual (error) variance. This
variance is the variation among experimental units treated alike;
replication is required to have "alike" units to compare. Of
course randomization is just as critical for validity of esti-
mation of variance as replication is. (2) The number of
replications greatly influences the sensitivity of the experiment. A comparison between treatments is made by comparing their
means. The number of observations from which the mean is estimated greatly influences the closeness of the observed mean to
the underlying true (population) mean; the more observations
that go into a mean, the closer the observed mean will get to
the true mean. Specifically, if σ^2 denotes the variance of
individual plants, then the variance of a mean is σ^2/n. This
occurs because observations above the mean average out with
observations below the mean; more observations give a greater
opportunity for this averaging out to function more completely.
(3) More replication increases the precision with which the
experimental (residual) variance (σ^2) is estimated. Specifically
increased replication gives more degrees of freedom to estimate
σ^2. Even a cursory examination of a table of significant values

of t or F shows how important it is to increase the degrees of
freedom associated with the estimate of σ^2, at least up to 30.

Two final thoughts on experimental design: statistical
procedures offer no substitute for careful planning and execution
of an experiment. Instead, statistics offers the tools of
experimental design to organize a study to minimize the impact of
known environmental variability and provide accurate estimates of
population parameters. Secondly, this merely highlights the
principles of experimental design. The next section gives some
specific illustrations of these principles in growth chamber
studies.

ILLUSTRATIONS

It is important to distinguish between variation within
growth chambers and variation between or among growth chambers in
designing experiments. When treatments are applied to an individual or group of plants and a single growth chamber is used as a
standard environment, then the investigator may only be concerned
with the within chamber variation. In this case the individual
or group of plants is the experimental unit and replication and
randomization should occur at this experimental unit level. As a
minimum, the experimental units should be randomized within the
chamber (Fig. 1). To account for within chamber variation and to
increase the sensitivity for detecting treatment differences, it
is suggested that small square blocks be used in growth chamber
studies (Hammer and Langhans, 1972; Lee, 1977; Rawlings, 1979).
Randomization of treatments should still occur within blocks
(Fig. 2). Lee (1977) and Rawlings (1979) found consistent
patterns of growth over the area of a chamber in different trials
(time) with the same plant material. However, the patterns were
different among chambers of the same or different types. This
could be used to suggest irregular shaped blocks for areas of
uniform growth or covariate analysis to account for the patterns.

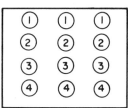

AVOID
no randomization

MINIMAL DESIGN
randomization

4 treatments (numbers)
3 replications

4 treatments (numbers)
3 replications

SYSTEMATIC DESIGN

No appropriate analysis of
variance exists

COMPLETELY RANDOM DESIGN

Source	*Degrees of Freedom*	*E(MS)*
Mean	*1*	
Treatments (nitrogen)	*3*	$\sigma_C^2 + Nitrogen$
Residual	*8*	σ_C^2

FIGURE 1. The experimental unit represented by the circles
has been replicated three times. Within chamber variation requires
randomization. The completely random design is appropriate if
there is no known pattern to the within chamber variation. It is
less precise than the design shown in Figure 2 if the within
chamber variation has a known pattern.

Covariate analysis would use the measured plant response (i.e. growth) at a specific location in the chamber from a previously conducted uniformity trial to "correct" the response of future studies. However, one would need to be sure that the growth pattern within a chamber did not change with time or experimental material before irregular shaped blocks or covariate analysis are used. For example, changing the numbers of plants and/or experimental apparatus in the chamber would likely change the growth pattern within that chamber, thus making covariate analysis useless. Lee (1977) and Rawlings (1979) also suggested that investigators should be careful not to overlook the non-homogeneity of variance within different types of chambers. For example, the variance may be different in a walk-in than in a reach-in chamber.

The last and probably most important point to be made is that replication *must* occur over chambers (of the same type) or with the same chamber over time if the chamber is the experimental unit; for example, if different levels of temperature, irradiance, or carbon dioxide are the treatments. Lee (1977) and Rawlings (1979) found that in most cases a within experiment (chamber) estimate of residual variability can badly underestimate the between experimental unit (chamber) estimate of residual variability by a very large amount. This simply means that when using separate chambers for the different treatments the experimenter may have differences due to the chambers used, and "see" this as a result of the applied treatment.

The best way to discuss the general principles of design and analysis when the chamber is an experimental unit is with one illustration approached in several ways. In the general case, the researcher may have one set of treatments utilizing several chambers and another set of treatments within each of the chambers. Blocks over time can be used at the chamber level; similarly, blocks within chambers account for within chamber variation. Suppose for example, air temperatures of 20 and 25°C are the main

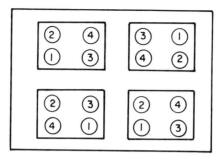

4 blocks

4 treatments (nitrogen level)

Source	Degrees of Freedom	E(MS)
Mean	1	
Blocks within chamber	3	$\sigma_{\mathcal{C}}^{2} + 4\sigma_{B}^{2}$
Treatments (nitrogen)	3	$\sigma_{\mathcal{C}}^{2} + Nitrogen$
Residual	9	σ_{C}^{2}

FIGURE 2. The experimental units represented by the circles have been arranged in small square blocks to account for within chamber variation.

plot (chamber) treatments and 4 nitrogen treatments are the within chamber treatments. If only one chamber is available to the investigator, the study could be run four times, 2 times at each temperature (Fig. 3). In this case the investigator will get an unbiased look at the differences in response at the two temperatures. The response may contain a treatment and chamber component; however, the chamber component should be nearly constant for each run and thus subtracts out the comparison between response. If such a design were used, the temperatures should be randomly allocated to runs and the blocks should be located in the same position within the chamber for each run for the ANOVA table in Fig. 3 to be used. If chamber blocks are not in the same position, the variances or Blocks within chamber and Error B cannot be

Time I 25° C

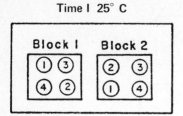

Time II 20° C

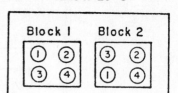

Time III 25° C

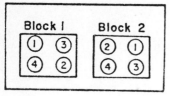

Time IV 20° C

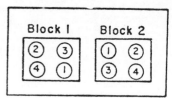

(The numbers in the circles represent
4 different nitrogen treatments)

Source	Degrees of Freedom	E(MS)
Mean	1	
Time (Blocks)	3	
Air temperature	1	$\sigma^2_C + 4\sigma^2_B + 8\sigma^2_A$ Temperature
Error A	2	$\sigma^2_C + 4\sigma^2_B + 8\sigma^2_A$
Blocks within chamber	1	$\sigma^2_C + 4\sigma^2_B +$ Position
Error B	3	$\sigma^2_C + 4\sigma^2_B$
Subplots within blocks	24	
Nitrogen	3	$\sigma^2_C +$ Nitrogen
Nitrogen X air temperature	3	$\sigma^2_C +$ Interaction
Error C	18	σ^2_C

FIGURE 3. When one chamber is used to study two temperatures blocked over time, differences in the response at the different temperatures will be unbiased comparisons. However, each response will be confounded with a chamber effect.

separated in the ANOVA table. Nevertheless blocking within chambers would still be important. There would be no change in the
rest of the analysis.

If two chambers were used, for the example, probably the
minimal design would be to block over time for 4 runs (Fig. 4).
In this case, the analysis of main plots would be a 2 x 4 latin
rectangle. Note the cross-over of temperature and chamber each
time and that the block effect within chamber A and B are
different. Again if the position of the block within a chamber
changes each time, Blocks within chamber and Error B cannot be
separated. In the ANOVA table Figure 4, several things become
very clear. Most of the power for testing differences is associated with the treatments within chambers. Differences in
response to air temperature would need to be very large in order
to be detected with such small degrees of freedom. One may want
to test Error A with Error B and pool them if appropriate (see
Bancroft, 1964).

If in the previous case only two time blocks were used, the
interpretation of results would be very difficult (Fig. 5). An
estimate of Error A would not be available for testing a temperature effect. The mean square for chamber might serve as an error
term in this case; however, with only one degree of freedom and
probably a chamber effect present, the test would be so conserva-
tive that significance would be almost impossible. Again note the
cross-over of temperature and chamber and the degrees of freedom
associated with Error C.

And finally we get to the worst case of confounding which is
probably the one many investigators have faced. If two chambers
are operated at two temperatures at one time there is no way
to test the temperature effect. It is confounded with chamber
effect and has no appropriate error term. The analysis of
variance in Figure 6 shows why a misleading result may occur
when comparing two chambers with different temperatures using

Time I (Block I)

Chamber A 20° C

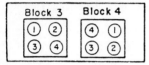

Chamber B 25° C

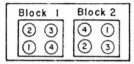

Time 3 (Block III)

Chamber A 20° C

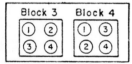

Chamber B 25° C

Time 2 (Block II)

Chamber A 25° C

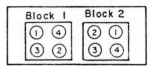

Chamber B 20° C

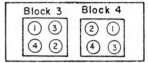

Time 4 (Block IV)

Chamber A 25° C

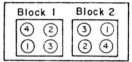

Chamber B 20° C

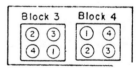

(The numbers in the circles represent 4 different nitrogen treatments)

Source	Degrees of Freedom	E(MS)
Mean	1	
Main Plot	7	
Time (Blocks)	3	$\sigma_C^2 + 4\sigma_B^2 + 8\sigma_A^2 + 16\sigma_T^2$
Chamber	1	$\sigma_C^2 + 4\sigma_B^2 + 8\sigma_A^2 + \text{Chamber}$
Air temperature	1	$\sigma_C^2 + 4\sigma_B^2 + 8\sigma_A^2 + \text{Temperature}$
Error A	2	$\sigma_C^2 + 4\sigma_B^2 + 8\sigma_A^2$
Blocks within a chamber	2	$\sigma_C^2 + 4\sigma_B^2 + \text{Position}$
Error B	6	$\sigma_C^2 + 4\sigma_B^2$
Subplots within blocks	48	
Nitrogen	3	$\sigma^2 + \text{Nitrogen}$
Nitrogen × air temperature	3	$\sigma_C^2 + \text{Interaction}$
Error C	42	σ_C^2

FIGURE 4. This would be the minimal design needed for unconfounded, unbiased results in the example discussed. The main plots are a 2 x 4 latin square design.

Time I (Block I)

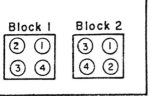

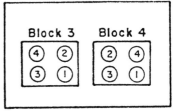

Time 2 (Block II)

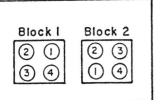

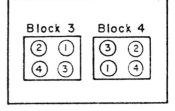

(The numbers in the circles represent 4 different nitrogen treatments)

Source	Degrees of Freedom	E(MS)
Mean	1	
Main plot	3	
Time (Block)	1	$\sigma_C^2 + 4\sigma_B^2 + 8\sigma_A^2 + 16\sigma_T^2$
Chamber	1	$\sigma_C^2 + 4\sigma_B^2 + 8\sigma_A^2 + \text{Chamber}$
Air temperature	1	$\sigma_C^2 + 4\sigma_B^2 + 8\sigma_A^2 + \text{Temperature}$
Error A	0	Not available
Blocks within chamber	2	$\sigma_C^2 + 4\sigma_B^2 + \text{Position}$
Error B	2	$\sigma_C^2 + 4\sigma_B^2$
Subplots within blocks	24	
Nitrogen	3	$\sigma_C^2 + \text{Nitrogen}$
Nitrogen X air temperature	3	$\sigma_C^2 + \text{Interaction}$
Error C	18	σ_C^2

FIGURE 5. In this case the main plot treatment was blocked twice over time. Although we can estimate the effect of time, chamber, and temperature, there is no estimate of error A with which to make a test.

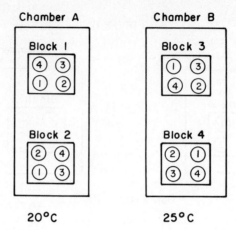

FIGURE 6. *In this case the main plot has not been replicated. There is no estimate of error A. Temperature and chamber effect cannot be separated and a test for temperature differences does not exist.*

Source	Degrees of Freedom	E(MS)
Mean	1	
Main plots	1	
Temperature ⎫		
Chamber ⎬	1	$\sigma_C^2 + 4\sigma_B^2 + 8\sigma_A^2 +$ Temp/Chamber
Error A ⎭	0	Not available
Blocks within a chamber	2	$\sigma_C^2 + 4\sigma_B^2 +$ Position
Error B	0	Not available
Subplots within blocks	12	$\sigma_C^2 +$ Nitrogen
Nitrogen	3	
Nitrogen x air temperature	3	$\sigma_C^2 +$ Interaction
Error C	6	σ_C^2

within chamber variance. The expected mean square, E(MS) for temperature, contains not only within chamber variance (σ_C^2) and the temperature effect, but it also contains $4\sigma_B^2 + 8\sigma_A^2 +$ chamber effects. Thus if error C is used to test for a temperature effect, a significant F test may tell us nothing about temperature. Significance could be due to temperature effect, chamber effect, chamber to chamber variation (σ_A^2), and within chamber environmental variation (σ_B^2). Any combination of other factors could appear to be temperature effects without any temperature effect existing. This is a real example of "seeing" differences that may not in fact be real.

In some cases confounding cannot be eliminated (e.g. the requirement of special lamp fixtures in a chamber when comparing different lamp types) and the investigator has little choice except to be aware of the confounding of treatment and chamber and report results accordingly. However, it is very important to repeat the treatments over time in this case.

These examples should show the importance of spending time with a statistician in the planning stages of each growth chamber experiment. Some will argue that growth chamber space is much too expensive for the amount of replication and blocking we have suggested here. Is it not cheaper to conduct one well planned and thus interpretable experiment than many poorly planned experiments that defy useful interpretation because of major confounding?

REFERENCES

Bancroft, T. A. (1964). Analysis and inference for incompletely specified models involving the use of preliminary test(s) of significance. *Biometrics* 20, 427-442.

Carlson, G. E., Motter, G. A., and Sprague, V. G. (1964). Uniformity of light distribution and plant growth in controlled-environment chambers. *Agron. J.* 56, 242-243.

Cochran, W. G., and Cox, G. M. (1957). "Experimental Designs." Wiley, New York.

Collip, H. F., and Acock, B. (1967). Variation in plant growth within and between growth cabinets. Report of the School of Agriculture, University of Nottingham: 81-87.

Federer, W. T. (1955). "Experimental Design." Macmillan, New York.

Fisher, R. A. (1960). "The Design of Experiments." 7th ed. Oliver and Boyd, London.

Gentner, W. A. (1967). Maintenance and use of controlled environment chambers. *Weeds 15,* 312-316.

Hammer, P. A., and Langhans, R. W. (1972). Experimental design consideration for growth chamber studies. *Hortscience 7,* 481-483.

Hammer, P. A., Tibbitts, T. W., Langhans, R. W., and McFarlane, J. C. (1978). Base-line growth studies of "Grand Rapids" lettuce in controlled environments. *J. Amer. Soc. Hort. Sci. 103,* 649-655.

Hruschka, H. W., and Koch, E. J. (1964). A reason for randomization within controlled environment chambers. *J. Amer. Soc. Hort. Sci. 85,* 677-684.

Kalbfleisch, W. (1963). Artificial light for plant growth. *In* "Engineering Aspects of Environmental Control for Plant Growth", pp. 159-174. Proc. Symp. Sept. 1-5, 1962, CSIRO, Melbourne, Australia.

Kempthorne, O. (1952). "The Design and Analysis of Experiments." Wiley, New York.

Knievel, D. P. (1973). Temperature variation within and between rooting media in plant growth chambers. *Agron J. 65,* 398-399.

Lee, C. S. (1977). Uniformity studies with soybeans at the North Carolina State University Phytotron. Thesis Series No. 1153. Institute of Statistics. North Carolina State University, Raleigh, North Carolina.

Measures, M., Weinberger, P., and Baer, H. (1973). Variability
of plant growth within controlled-environment chambers as
related to temperature and light distribution. *Can. J. Plant
Sci. 53,* 215-220.

Mitchell, C. A., Severson, C. J., Wott, J. A., and Hammer, P. A.
(1975). Seismomorphogenic regulation of plant growth.
J. Amer. Soc. Hort. Sci. 100, 161-165.

Rawlings, J. O. (1979). Personal Communications. Department of
Statistics, North Carolina State University, Raleigh, North
Carolina.

Tibbitts, T. W., McFarlane, J. C., Krizek, D. T., Berry, W. L.,
Hammer, P. A., Hodgson, R. H., and Langhans, R. W. (1977).
Contaminants in plant growth chambers. *Hortscience 12,*
310-311.

Tibbitts, T. W., McFarlane, J. C., Krizek, D. T., Berry, W. L.,
Hammer, P. A., Langhans, R. W., Larson, R. A., and Ormrod,
D. P. (1976). Radiation environment of growth chambers.
J. Amer. Soc. Hort. Sci. 101, 164-170.

*Purdue University, Agricultural Experiment Station, Journal
Article number 7682, and Journal Article number 727 of the
Agricultural Experiment Station, New Mexico State University,
Las Cruces, New Mexico.*

PRECISION AND REPLICATION:　DISCUSSION

KOLLER:　In view of the lack of uniformity in growth chambers, would moving plants around daily on a random basis improve reproducibility or reduce the error?

HAMMER:　There are several concerns with this procedure.　The vibration from moving could add to the lack of uniformity, particularly if all the plants were not moved uniformly.　It would also spread the variability over the experimental material, which would increase instead of reduce the error.　Blocking will allow you to estimate this lack of uniformity within the chamber and increase sensitivity to treatment differences.　However, moving the plants randomly would be better than lining the treatments up in rows.[1]

TIBBITTS:　I would like to ask Kostkowski to react to our regional committee's (NCR-101) planned project to calibrate a quantum meter and pyranometer at the Bureau of Standards and then distribute these meters to different laboratories to calibrate their instruments under their particular combination of

[1]After additional thought and discussions with Urquhart, this answer should be modified.　If the plants were moved without significant vibration, and each plant was rotated into each position within the chamber several times during the course of a study, the block effect we described would disappear, and residual variability would be unaffected.　The only possible gain would be a modest increase in residual degrees of freedom.　But the suggested procedure could introduce some unanticipated biases resulting from patterns in which the plants are moved.

lamps. What accuracy in radiation measurements can we obtain with such a procedure?

KOSTKOWSKI: Without some lengthy calculations, an estimate of this accuracy will be only a rough guess. And to perform the calculations, specific information is needed about the instruments and the radiation fields to be measured. For example one needs the relative spectral and geometrical responsivity of the instruments and the uncertainties of all these quantities. Incidently NBS is not now set up to determine the responsivity function of these instruments so that one of the laboratories involved in the program or a commercial laboratory would have to perform this measurement. Ideally, each of the laboratories should also measure the relative spectral distributions in the chambers (and geometrical distribution if the instruments are not cosine corrected). It is possible to use manufacturers' "average" data for the various lamps and for the radiometers. Then one would only have to make a measurement of a tungsten lamp standard and calculate or estimate the effect of the chamber on the radiation emitted by the lamps. My *guess* is that your errors would be greater than 20% if you did this. If instead of a tungsten lamp, one had a standard that was similar to the radiation sources being measured, the errors would be less. NBS could provide the tungsten standard and possibly a cool white fluorescent standard.

The situation should be much better, relative to what NBS can offer, by the spring of 1980. Then, we expect to have a facility for determining the relative spectral responsivity of a radiometer from the UV to 800 nm and a portable, calibrated, easy-to-use spectroradiometer that would be available for rent and could be used to determine the spectral distribution of the radiation in the chambers, at least from the UV to 800 nm. With these measurements your uncertainties with the quantum meter should decrease to 5% or less.

PALLAS: We cannot use an Eppley radiometer for measurements of black body radiation above 2800 nm. You may want to consider an instrument like the Beckman-Whitley.

Recently I had a discussion with ISCO concerning the response of their ISCO spectroradiometer to sunlight. I was told that they were not going to produce the instrument any more because of such a small market.

McFARLANE: What will be the spectral band and rental cost of the spectroradiometer from the National Bureau of Standards?

KOSTKOWSKI: We do not know what the rent will be yet. The instrument is a double monochromator that will cover radiation from 200-800 nm. Incidentally, our current routine capability at the National Bureau of Standards does not include broad band source standards. We had a standard that covered total irradiance but only sold about one every four years. We could not afford to maintain those standards with state-of-the-art accuracy, so our presently available standards are limited to spectral standards that go out to 2 1/2 microns.

NORTON: A number of commercial laboratories around the country perform the service of calibrating instruments. Does the National Bureau of Standards certify these laboratories, and can we have confidence in using them to calibrate our instruments?

KOSTKOWSKI: The National Bureau of Standards does not certify any of the commercial standards laboratories. It would be a tremendous task to prove the reliability of these laboratories and to continually maintain this proof, particularly for the wide range of calibrations in which the radiometric community is interested. We do not have the resources to do this.

ORMROD: We are not proposing guidelines for experimental designs in growth chambers. Would Hammer make some recommendations on a manual or reference text that would be useful to ourselves, technicians, and graduate students in designing growth chamber experiments?

HAMMER: That is a difficult question to answer. All of us have seen very few textbook experiments. Each experiment is unique and should be approached in this way. I would suggest you obtain copies of the references listed in our paper and visit a statistician when planning a growth chamber study. That is why he is in business. A good example along this line is trying to fit data to the package computer programs instead of designing the study to answer a question of interest to the investigator. This does not imply that the investigator should overlook the topics of randomization, replication, and blocking in growth chamber studies. They should be carefully considered and the investigator should understand the possible errors when certain assumptions are made.

TISCHNER: Along with the discussion of accuracy and precision in growth chambers, we should mention reliability. When a chamber breaks down during an experiment, we many times lose the entire study. I think we should deal more with this problem in the future. We may demand that manufacturers guarantee the reliability of their chambers in the future.

WURM: As a representative of a company that makes the quantum sensor, I appreciate your inviting Kostkowski and the comments he made concerning the measurement of optical radiation. Our own experience has been that you cannot assume anything and indeed the errors being discussed are real. We have patterned our calibration set-up after National Research Council in Canada, which I assume is similar to the one at the National Bureau of Standards. We use a tungsten source and feel that it is the most accurate and repeatable measurement we make. Under some lamps, we can have a 10% error rate, but the number of different lamps we can measure under is small.

I also appreciate the emphasis on recalibration. We have found many people who have used our equipment for four years and still assume nothing has changed. We would recommend, at the very minimum, that the instrument should be recalibrated every two years and probably more often if you want to assume the instrument has not drifted.

FRANK: You indicated blocking over time instead of chambers. Is this because of unreliability of chambers?

HAMMER: No, if you look at the estimates of variability from the two sources, the chamber variance is smaller than the time variance. So, if you can only block over one factor, choose the largest source of "unwanted" variability to block over.

McFARLANE: It seems that with the new rapid data collection systems we are talking about collecting a great deal of environmental measurement data. We as a group of scientests should push our editors to accept these data in our publications.

INTERACTIONS OF ENVIRONMENTAL FACTORS

Wade L. Berry

Laboratory of Nuclear Medicine and Radiation Biology
University of California
Los Angeles, California

Albert Ulrich

Department of Soils and Plant Nutrition
University of California
Berkeley, California

INTRODUCTION

Many controlled environment studies have been designed to very carefully control and monitor one or at most a few environmental factors while paying relatively little attention to others. It is becoming increasingly important, however, to carefully control a number of additional factors in order to assess the impact on plants of numerous interactions among such factors. Such interactions may be synergistic, additive, independent, or antagonistic.

This paper will discuss interactions among environmental factors and will emphasize interactions of mineral suppy with water, temperature, air pollution, light intensity, and nutritional preconditioning.

WATER

The environmental parameter that probably is most closely
associated with mineral nutrition is water. The thirteen mineral
nutrients required by plants are normally acquired from the soil
solution in a soluble form. This water is the major medium by
which these mineral nutrients are supplied to plants. We are all
aware of the importance of water of good quality for irrigating
crops. In fact, numerous governmental agencies have established
water quality standards for irrigation. The Water Quality Criteria Committee of the Department of Interior (U.S. Dept. of Interior, 1972), has been much involved in establishing standards.
In established criteria for water quality two concentrations for
many trace elements are given depending on the soil texture. It
is assumed that the parameter of primary concern is the cation
exchange capacity(CEC). Thus, even at the crop production level,
significant interactions have been shown between rooting media
and water quality on plant response. An important question is
how such water quality criteria, which were developed for field
use, can be applied to controlled environment conditions. Direct
application of such criteria is limited to a few situations where
the controlled environments were chosen to reflect field conditions. However, if the principles and processes used in developing the criteria are kept in mind they can often be applied to
individual controlled-environment studies.

In establishing standards for quality of irrigation water
primary attention has been given to specific toxicity of some of
the heavy metals. Separate criteria have been established for
heavy and light soils. This is because for many mineral elements
the total amount of nutrients in the soil is divided into at
least two fractions, only one of which is readily available for
uptake by plants (Lindsay and Norvell, 1978). This is illustrated
by the large number of extractants used to evaluate the nutrient
status of different soils. Usually all of the nutrients added to
light-textured soil are readily available to plants; in heavier
soils more of the added nutrients are unavailable. Hence a

potentially toxic element can induce injury at a lower concentration in a light soil than in a heavy soil.

Unfortunately, controlled environment experiments have often been designed to make added nutrients either very available or very unavailable. For example, in experiments with solution cultures all or almost all of the added nutrient elements are readily available to plants. In experiments using artificial rooting media, such as peat or vermiculite, a large proportion of the added heavy metal nutrients may not be readily available. Hence such media have characteristics of heavy soils.

An interaction between water and nutrition may occur when a nutrient solution is recycled for an extended period of time. Presently much controlled environment research is restricted to relatively short experimental periods (1-3 months). In addition it is quite common to replace the entire nutrient solution periodically. When the duration of experiments is increased and the nutrient solution is recycled rather than replaced, small enrichment factors of trace elements in the system, or additions, can quickly accumulate to toxic concentrations. This has been confirmed with both aggregate culture and nutrient film techniques in commercial hydroponics operations. If a system has a nutrient solution reservoir of 8 liters per plant, as commonly used in commercial production of tomatoes by hydroponics, and evapotranspiration is 2-4 liters per day, the solution is completely replaced by make-up water in 2 to 4 days. Very small concentrations of trace elements in the make-up water could therefore quickly build up to potentially toxic concentrations (Berry, 1978).

TEMPERATURE

As already mentioned temperature has a very direct impact on plant growth. The range of temperature in which plants grow is

very limited and generally, as temperature increases, the growth
rate of plants increases until the upper threshold of temperature
tolerance is reached (Ulrich, 1952a). There also are several secondary effects of temperature on plant growth which can be illustrated by a phosphate-temperature interaction which often is economically important in production agriculture (Sipitanos and Ulrich, 1970). Responses of plants in the field to added phosphorus
consist of large increases in vegetative growth early in the growing season and a significant increase in yield. The response to
phosphate occurs even through classical soil analysis shows that
soil phosphorus is adequate, and plant analysis at harvest indicates that plants are well supplied with phosphorus. For example,
in field plot experiments with sugar beets, leaf samples taken at
regular intervals during the growing season have demonstrated this
unlikely phosphorus response. Early season petiole phosphate-phosphorus values were well above the critical concentration of 750 ppm
(dry wt) for all plants on phosphated plots, whereas values for
plants on non-phosphated plots were well below this concentration.
However, only three weeks later, phosphate-phosphorus of non-phosphated plants was very much higher and was as high or slightly
higher than in plants on phosphated plots.

In order to sort out the possible factors responsible for this
response, experiments were conducted including one in which the
rates of phosphate fertilization and soil temperature were studied.
This experiment showed that all plants on untreated soils were
deficient in phosphorus initially, regardless of soil temperature
(15, 20, 25, or 30°C). When the soil temperature was 30°C, phosphorus deficiency lasted only a few weeks at the beginning of the
experiment. However, when the soil temperature was 20°C the
plants remained deficient in phosphorus until they were harvested
at the age of seven weeks. Yields of plants on untreated soils,
as compared to those on phosphate-fertilized soils, were 6% at
15°C, 16% at 20°C, 31% at 25°C, and 40% at 30°C., emphasizing a
strong interaction between phosphate nutrition and temperature.

AIR POLLUTION

There is growing concern with interactions of air pollution
and other environmental factors. Initial symptoms of smog damage
resemble those of potassium deficiency and the two could be confused unless the nutritional history of the experimental plants
is known. Identification of smog damage to plants was a serious
early problem in the phytotron at Pasadena. When nutritional experiments were conducted, symptoms of leaf injury were often observed and did not appear to be associated with nutritional treatments. Subsequently it was shown that the injury symptoms could
be reduced by filtering the air, and they could be enhanced by
vapors given off by plastic or other organic vapors. This eventually led to inclusion of activated carbon filters in the air supply system of the phytotron and in many research greenhouses.

Air pollutants may also be introduced inadvertently with the
hardware necessary to maintain controlled environmental conditions.
Tibbitts *et al.* (1977) gave examples of toxic effects of solvents
used in construction of growth chambers. Such solvents may escape
slowly from glues or sealants used in construction and accumulate
to toxic levels in growth chambers because of the constant recy-
cling of air within the chamber. Similarly mercury may accumulate
in growth chambers and cause severe plant damage (Waldron and
Terry, 1975). A few drops of mercury which escape from a broken
thermometer are sufficient to cause severe damage to plants even
in large growth chambers. Damage to leaves may appear within
hours and the chamber air will continue to injure plants for a
long time unless stringent measures are taken to clean up the
spill. The extreme sensitivity of plants to mercury vapor will
cause severe damage to sugar beets at concentrations less than
0.1 mg m^{-2}, which is below the detection limit of a Lemaire
ultraviolet absorption meter. This dose level is also an order
of magnitude lower than that which is toxic to rabbits after 2 to
3 weeks of exposure.

LIGHT INTENSITY

There are many interactions of light intensity and other factors. In fact, characterizing the primary action of light on
plants is extremely difficult. Historically, controlled environments were not used in plant research until radiation of a minimum of 80 μE m^{-2} s^{-1} during long days was possible from artificial
light sources. Interaction of light with nutrition is generally
not direct but channeled through one of the more primary effects
such as that on temperature or growth rate.

Experiments with sugar beets showed that a threshold of light
is required to meet minimal functional needs before normal plant
growth can proceed. This threshold is about 80 μE m^{-2} s^{-1} for
sugar beets during a 16 hr photoperiod; additional light leads to
increased growth. The upper limit for light utilization in this
plant is apparently near 640 μE m^{-2} s^{-1} for a 20-hour day. How
increased light energy is utilized by plants depends on a number
of factors. For example as the amount of light received by sugar
beets is increased by extending the photoperiod, top growth is not
appreciably altered. Most increase in growth is due to increased
beet size. The extent to which increased light can influence
growth rate also depends on temperature. At very high (>30°C) or
very low (< 17°C) temperatures, increased light is not efficiently
utilized (Ohki and Ulrich, 1973). When the nitrogen status of the
plant is favorable, additional light is used in increasing beet
size. If the plant is very slightly deficient in nitrogen it is
still able to utilize light but sugar production is favored over
increase in beet size (Ulrich, 1952b; Ulrich 1954). Hence the
nitrogen status of the plant can be used as a switch to produce
plants with large beets or with high sugar, but beet growth and
sugar accumulation do not occur at the same time. In order to produce large beets with high sugar content the two processes must
occur in the proper sequence.

NUTRITIONAL HISTORY OF PLANTS

The fact that growth of plants is influenced by their past history should be taken into consideration in planning controlled environment research. Past histroy can be important even when plants
are started from seeds. Preconditioning of plants is particularly
important when seedlings from outside sources are used in controlled environment experiments (Ohki and Ulrich, 1975). The term
"antecedent nutrition" has been used to indicate that the nutritional history of test plants influences plant responses to current
environmental conditions. Ohki and Ulrich related potassium absorption in barley to the status of other nutrients in the plants.
Absorption of potassium depended greatly on the nutrient and salt
status of the plant immediately prior to the time of absorption.
Potassium-or nitrogen- stressed plants rapidly absorbed potassium
whereas plants low in other nutrients were less effective in absorbing potassium. The classical studies of Hoagland and Broyer
on salt absorption were always performed on plants referred to as
low salt plants. These plants had high rates of salt uptake when
compared with plants with a higher level of nutrition. Such studies indicate that plants have some type of a feedback loop by
which plant response to given environmental conditions depends on
their past histroy. This response extends beyond mineral absorption and can influence growth and development of plants over their
entire life cycle. For example, if seedling lettuce is stressed
for nutrients only during the first two weeks of growth, subsequent
growth can be reduced as much as 60%.

Another example of the effect of the nutrient history of plants
is related to the trace element content of seeds. Large-seeded
plants may contain sufficient amounts of some of micronutrients,
such as molybdenum, to carry them through their entire life cycle.
Thus it would appear unnecessary to supply molybdenum from the environment to such plants. It is more common, however, to find

seeds of small seeded plants with such a low content of macronutrients that they require an external supply of such nutrients as
nitrogen and potassium within the first three days after imbibi-
tion. When seeds of such plants are germinated in sand with dis-
tilled water they may already be nutrient-deficient by the time
seedlings emerge.

Inasmuch as one of the primary purposes of this conference is
to propose guidelines for measuring and reporting environmental
factors the reporting procedures for mineral nutrients in nutrient solution systems should be considered. It would be consistent
with the emphasis given to characterizing and monitoring other
environmental factors to also report mineral nutrient concentrations as moles per unit volume. Both macro- and micronutrients
would be amenable to this form of reporting. The macronutrients
could be reported as mM/l and the micronutrients as μM/l. However,
in reporting nutrient supply in nutrient solution systems it is
also necessary to report volume of solution per plant per unit of
time. The solution sample for analysis should be taken from the
immediate vicinity of the roots and, at a minimum, sampled at the
beginning and end of an experiment.

REFERENCES

Berry, W. L. (1978). Comparative toxicity of VO_4^-, $CrO_4^=$, Mn^{++},
    Co^{++}, Ni^{++}, Cu^{++}, Zn^{++}, and Cd^{++} to lettuce seedlings. Envi-
    ron. Chemistry and Cycling Processes Symposia (D. C. Adrano
    and I. Brisbin, eds.), pp. 582-589. ERDA Symp. Series Conf.
    760429.
Lindsay, W. L., and Norvell, W. A. (1978). Development of DTPA
    soil test for zinc, iron, manganese and copper. *Soil Sci.
    Soc. Amer. J. 42,*421-428.
Ohki, T., and Ulrich, A. (1973). Sugarbeet growth and development
    under controlled climatic conditions with reference to night

temperatures. *Proc. Amer. Soc. Sugar Beet Technol. 17*,270–
279.

Ohki, T., and Ulrich, A. (1975). Potassium absorption by excised
barley roots in relation to antecedent K, P, N, and Ca nutrition. *Crop Sci. 15*,7–10.

Sipitanos, K. M., and Ulrich, A. (1970). The influence of root
zone temperature on phosphorus nutrition of sugarbeet seedlings. *Proc. Amer. Soc. Sugar Beet Technol. 16*,408–421.

Tibbitts, T. W., McFarlane, T. C., Krizek, D. T., Berry, W. L.,
Hammer, P. A. Hodgson, R. H., and Langhans, R. W. (1977). Contaminants in plant growth chambers. *Hortscience 12*,310–311.

Ulrich, A. (1952a). The influence of temperature and light factors
on the growth and development of sugar beets in controlled
climatic environments. *Agron. J. 44*,66–73.

Ulrich, A. (1952b). Physiological basis for assessing the nutritional requirements of plants. *Annu. Rev. Plant Physiol. 3,*
207–228.

Ulrich, A. (1954). Growth and development of sugar beet plants
at two nitrogen levels in a controlled temperature greenhouse.
Proc. Amer. Soc. Sugar Beet Technol. 8,325–338.

U.S. Dept. of Interior. (1972). Report of the Committee on Water
Quality Criteria, p. 152. Washington, D.C.

Waldron, L. J., and Terry, N. (1975). Effect of mercury vapor on
sugar beets. *J. Environ. Qual. 4*,58–60.

INTERACTIONS OF ENVIRONMENTAL FACTORS: DISCUSSION

TIBBITTS: Will plant nutritionists accept millimoles instead of milliequivalents and ppm units for nutrient concentrations?

BERRY: Because there is such a strong effort to obtain acceptance for SI units in all fields of science, I believe plant nutrition researchers will be happy to make the changes. However, production agriculturists may not be so willing to accept this change.

NORTON: Why is there objection to the use of ppm units which are so well established in production practise that it will be difficult to make a change?

BERRY: There is a lack of relationship of ppm to chemical activity. Micromoles are directly related to chemical activity for all elements. For example, elements bind to chelate on a mole to mole basis, not on a ppm basis.

KOZLOWSKI: The growth habit of the plant itself can interact with the environment. In woody plants of temperate zones there are basically two patterns of shoot growth, "fixed" growth and "free" growth. In species with fixed growth, shoot formation involves bud differentiation during one year (n) and rapid extension of the preformed shoot during the next year (n + 1). In such species shoot expansion often occurs in less than half of the frost-free season. In contrast, in species with free growth most elongation during year n + 1, continues much later into the summer, and includes not only expansion of the leaves that formed during the year n but also initiation and expansion of new leaf primordia. In species with fixed (predetermined) shoots, favorable environmental conditions during year n induce formation of large buds that will expand during a few weeks of year n + 1 into long shoots with many leaves. Hence, environmental conditions during the year of bud formation (n) usually determine shoot length more than conditions during year n + 1.

By comparision, in species exhibiting free growth, both shoot
length and number of leaves are influenced much more by environmental conditions during year n + 1.

KRAMER: There are several examples of prior treatment affecting subsequent plant growth. Environmental conditions under
which the seeds are produced can affect later growth. If absolute uniformity is desired in a population of plants it is probably better to grow seeds for these plants in controlled environments.

KOLLER: In *Amaranthus* the environmental conditions to which
the apical meristem is exposed prior to floral initiation affect
the seeds that are formed in a way that can influence the growth
of seedlings developing from such seeds.

TIBBITTS: There is an interesting interaction in many controlled environments that leads to the development of tumorous
growths on leaves and stems of plants. Lack of UV radiation and
high humidity encourages injury. In addition we think there is a
contaminate involved that is inactivated by UV radiation.

BERRY: When plants are grown in nutrient film, a very limited root system develops which requires that high humidity be
maintained. This is necessary to avoid high evaporative demands
that would exceed the capability of the limited root volume to
supply water.

The interaction of humidity with soil temperture should be
emphasized. When plants are grown in clay pots, evaporation and
hence soil cooling varies with the humidity of the air.

AIR MOVEMENT: GUIDELINES

Murray E. Duysen

Department of Botany
North Dakota State University
Fargo, North Dakota

Air movement facilitates the exchange of heat, carbon dioxide, and water vapor between plants and the bulk atmosphere by enhancing surface ventilation that removes the boundary layer of air (Grace, 1977). The rate of air flow near plant surfaces is dependent upon the air velocity, the pattern of air flow through the plant canopy, and morphological-anatomical features of the foliage.

Air movement is a critical environmental factor to measure since it influences gaseous and heat exchange rates as well as plant growth. In controlled environments, air movement interacts with radiation and humidity to influence plant temperature, evapotranspiration, and carbon dioxide exchange rates (Drake, 1967; Gates, 1976; Grace, 1977). The uniformity of temperature, humidity, and carbon dioxide is maintained in environmental enclosures by circulating conditioned air through the growing area, either upward, downward, or laterally (from side to side) (Downs, 1975). Thus, sufficient air movement is provided in chambers to permit maximum control of the environment (Solvason and Hutcheon, 1965) with a minimum of effects that could limit plant growth. Containers and plant canopies divert air flow patterns and alter air velocity in controlled environments (Downs, 1975). The measurement of air movement is complicated by the turbulent nature of flow which results in rapid fluctuations in direction and velocity at any one point within the chamber.

It is recommended that air velocity be measured at the top of the plant canopy in controlled environment enclosures, near the region of active plant growth. A sufficient number of areas over the canopy should be selected at random and successive readings taken at each location to ensure sampling the range of variation due to turbulence. The average air velocity and range in space should be reported in m sec^{-1}, the SI recommended unit. Since air movement can be influenced by plant size and leaf shape (Downs, 1975; Grace, 1977), measurements should be taken at the beginning and the end of the studies.

The ability to estimate air velocity accurately could be limited by the instrument selected to measure it. Recently, Downs (1975) and Krizek (1978) have reported on instruments that measure the air velocity in controlled environments. The anemometer selected should be sufficiently accurate and sensitive to monitor the low air velocities of controlled environment enclosures Single and multiple hot-wire anemometers have been suggested as well as certain mechanical anemometers (vane or cup). The type of air velocity meter used should be reported as well as the direction of air flow (ASHS Committee on Growth Chamber Environments, 1977). Measurements should be made with the chamber doors closed and with plants in the growing area (Downs, 1975).

REFERENCES

ASHS Committee on Growth Chamber Environments. (1977). Revised guidelines for reporting studies in controlled environment chambers. *Hortscience 12,* 309.

Downs, R. J. (1975). "Controlled Environments for Plant Research." Columbia University Press, New York.

Drake, B. G. (1967). Heat transfer studies in *Xanthium*. M.S. Thesis. Colorado State University, Fort Collins, Colorado.

Gates, D. M. (1976). Energy exchange and transpiration. *In* "Water and Plant Life" (O. Lange, L. Kappen, and E. Schulze, eds.), pp. 137–147. Springer-Verlag, New York.

Grace, J. (1977). "Plant Response to Wind." Academic Press, New York.

Krizek, D. T. (1978). Air movement. *In* "A Growth Chamber Manual: Environmental Control of Plants" (R. W. Langhans, ed.), pp. 107–116. Cornell University Press, Ithaca, New York.

Solvason, K. R., and Hutcheon, N. B. (1965). Principles in the design of cabinets for controlled environments. *In* "Humidity and Moisture" (A. Wexler, ed.), pp. 241–248. Reinhold, New York.

AIR MOVEMENT: DISCUSSION

JAFFE: Since much of my work involves effects of mechanical stimulation on plants and we have been studying effects of wind on plant growth and development, I would like to comment on wind and growth chambers. First, the effects of wind are very pronounced. In field experiments, over a range of temperature (6°C-29°C) and irrigation regimes, we see morphological changes in plants due to wind. Wind is a factor at rates from about 3 miles hr^{-1} or higher, which is enough to gently cause leaves to flutter. In growth chambers there are at least 3 factors that will cause mechanical effects on plants and one is wind. Turbulence is very important. As long as plants are in growth chambers there will be trubulence because the leaves flutter and cause turbulence. This will reduce laminar air flow. So when investigators take measurements above the leaf canopy they will not necessarily know the wind speed in the canopy, even at the very top. Another factor is vibration. Someone told me the other night that one of the fan blades of a growth chamber was bent and caused the growth chamber to vibrate. This will affect growth of plants. Finally, due to mechanical aber-
will affect growth of plants. Finally, due to mechanical aberration, there could be noise effects. In preliminary experiments we have found certain frequencies and certain decibel levels to have very pronounced effects on plant growth.

What can be concluded from all this? First, in nature there are always mechanical perturbations. We should have these in growth chambers. Investigators should not try to set up facilities without wind, vibration, and sound if they want something approaching natural conditions. The important thing is to know the magnitude of the mechanical perturbation. Measurements of wind speed or even vibrations should be made.

One other point, when large walk-in rooms are used, plants are carried or often wheeled in on carts. This procedure can set up vibrations. In 24 hours there can be up to 25-30% reduction in elongation of beans gently shaken for 5-10 seconds by running the hand through the foliage. This is bound to have an effect on results unless the control plants are treated similarly.

HELLMERS: Height growth of plants is reduced by 25%. What effect does moving plants have on dry weight increment?

JAFFE: Usually, there are pronounced effects on dry weight changes. It depends on the kind of stimulus: wind, shaking, or rubbing, and the intensity. In some cases, there is an increase in dry weight increment and in others a decrease.

HELLMERS: Are these effects additive? In all of our growth rooms and greenhouses we always keep the leaves moving by the air speed. We also move plants from chamber to chamber. Are we compounding our problems or does the fact that we move the leaves continuously compensate for moving the plants on carts?

JAFFE: The intensity of the stimulus is what is important. If there is enough of a stimulus due to plants blowing in the chamber you are probably saturating the effect. Moving the plants with a cart will not add to the effect. If there are short plants in some chambers with fans at the top moving air, it is not until the plants grow into the air stream that they begin blowing around. If the investigator is working with small plants and moves them, then there will be an effect. If plants have been blowing around in the growth chamber all along there probably will not be an additional effect from an added mechanical stimulus.

ORMROD: Mary Measures of Ottawa did her doctoral thesis about ten years ago on effects of sound on plants. I think she will be delighted to hear that you are finding effects. Such work needs additional study.

In growth chambers of the Canberra phytotron, I noticed that they lifted certain plant beds where they had vibration and put springs under them to minimize the problem. That might be a suggestion for those who have growth chambers in which there are vibrations in the beds.

HAMMER: We have found not only a 50% reduction in height growth of tomato plants stressed twice a day by shaking for 30 seconds on a gyratory shaker over a 35-day stress period, but also a reduction in node number from 21 to 17. There was also a reduction in fresh and dry weight. I would reinforce Jaffe's statement. If the investigator must move some plants, he should move all of them. If certain plants are sprayed with a chemical, control plants should be sprayed with solution lacking the test chemical.

JAFFE: If some of the plants in a growth chamber are treated to cause ethylene production, the ethylene will diffuse and affect the other plants in the same chamber. We have found this particularly true when treatments consisted of mechanical stimulation. Stresses from desiccation and chilling, result in ethylene production in the treated plants and unless the air is rapidly renewed, the control plants are affected also.

BUGBEE: We had the chamber with the bent fan and were able to correct the chamber vibration. Some chambers have fans and condensers below the plant area and have much more vibration than others that have external fans and condensers. I think we should raise the consciousness of people to chamber vibration by suggesting some simple measurement of it and placing it in the guidelines. How might we get an idea of chamber vibration for the guidelines without buying a great deal of expensive equipment?

SALISBURY: We sprayed tomato plants with water to reduce height growth because they get too long and leggy in winter in commercial practice. Water sprays reduced height increase by 40%, about the same amount as manual shaking. Spraying with a gentle mist instead of a hose that mechanically stressed the plants did not reduce height growth significantly. The important result was that dry weight increment and yields were

reduced.   In fact yields were reduced by 10-15%, too much for
commercial practice.   The investigator will have to decide
whether he prefers the shorter plants and reduced yields or long,
leggy plants and a slightly higher yield.

FORRESTER: We had a customer who was very concerned about
vibration.   We prepared the conditioning system as a completely
free-standing, isolated box that had a discharge and return duct.
The experimenter built his own enclosure and provided the air
from the conditioning system and passed it back.   This was a
way of isolating the sound from the enclosure.

In building a walk-in enclosure, one should arrange to build
on a concrete floor.   In this way it is much easier to isolate
vibrations from the plants on the carts, particularily if the
floor is properly prepared prior to the time the chamber is
installed.   Even retrofitting an installation is a possibility.
I have seen this done by jacking up the chamber, removing
the floor panels, and setting the chamber back on the floor.
After the equipment was isolated most of the vibration problems
were solved.

JAFFE: Since plants are exposed to wind in nature, there
should always be air moving plants in chambers to make sure there
will not be mechanical effects due to air movement that con-
found results.   The growth chambers should be designed or baffles
should be installed, so all plants will always be stimulated by
the air movement.   This will assure lack of variability between
plants that are mechanically stimulated and those that are not and
will assure that plants are growing more like those in the field.

Normally, in nature, plants are not exposed to vibrations.
There is an easy way of damping vibrations without buying
damping tables.   We place four tennis balls in the corners on
the shelves and place another shelf on top of the balls.   The
tennis balls act as damping agents and we get little vibration
of the plants.

TIBBITTS: What is the threshold air speed that provides maximum stimulation to plants? We need that information badly. Will it vary with different types of plants?

JAFFE: The only data I can give are for beans. We find the saturation point is reached at about 15-16 km hr^{-1}. That is a wind between a force 1 and 2, enough to make the plants wave gently. In fact, the beans we observed in the Biotron room were blowing enough to probably saturate the effect.

HELLMERS: Did you say 16 km hr^{-1}? You are talking about 10 miles hr^{-1}. That would really whip things around. We move air at about 1 mile hr^{-1}. That makes plants move.

JAFFE: That's true. If you have moving air in a chamber without any plants you would have a 1 or 2 mile hr^{-1} wind. With the plants imposed in the air stream, turbulence and a much higher wind speed are created.

KLUETER: I have a question on measuring the horizontal air flow. It seems that the top of the canopy is not sufficient. Could you comment on that?

DUYSEN: We are suggesting an average air movement over the top of the plant growing area. If the readings are taken at random, it should be possible to get that value. It should not make any difference if the air is moving up, down, or from side to side in the chamber.

FORRESTER: Let me destroy an old wives' tale about horizontal air flow. Remember there is a fixed heat load and a fixed amount of air moving in the box. The laws of physics tell how much heat will be transferred with that air. If you have laminar flow, that is going to determine what the maximum gradient will be in the box. We are not dealing with laminar flow when the air is moving horizontally. We do not want laminar flow in a growth chamber. The minute something is placed in the box, there is turbulent flow. The boxes have turbulent flow and they are going to backmix. There will not be gradients because the air comes in on one side and goes out the other.

Chambers do not have gradients if they have proper backmixing. Now, when the measurements are taken there will be gradients depending upon the shielding and design. We have built vertical and horizontal chambers and we get minimum gradients in them. The minute plants are placed in them, gradients are created and the air velocity is increased at some points.

HAMMER: We had a researcher set up a humidity gradient in a box that had air flowing horizontally. The gradient was created by the transpiring plants in the chamber. He did some experiments on differences in relative humidity with the horizontal flow.

FORRESTER: You can do that. If you were to have a 150 ft. min^{-1} air flow, the usually recommended rate, and an open canopy so there is backmixing, there will not be gradients. But if the air flow is reduced by essentially using the plants as an air filter, laminar flow is induced and gradients are created.

KRIZEK: Currier at the University of California, Davis, observed some years ago the induction of callose formation as a result of moving plants on a cart. Thus there are many phenomena, in addition to some of the ethylene-mediated responses pointed out here, that need to be considered in conducting controlled environment studies.

FORRESTER: The anemometer that is on display at this conference is extremely directional sensitive. If the position is changed 5-10°, entirely different readings are obtained. A head is made with a protective cap that gives laminar flow over the anemometer wire and it is much easier to use although larger in size. Measurements taken with the caps in place are much more reproducible.

Air velocity differences are minimized in the box if air is moved from one entire side of the chamber to the opposite side, either horizontally or vertically.

PALLAS: We haven't talked about one of the major reasons for air flow and that is to reduce boundary layer resistance.

Most species need 1 to 2 miles hr^{-1} over the leaves for maximal growth.

BAILEY: A significant difference between upward and downward air flow in a chamber is that downward air flow causes turbulence below the pots because the pot line is the area of greatest restriction to air flow. Upward air flow produces turbulence above the containers.

BUGBEE: Would it be reasonable to include in the guidelines the use of an inexpensive and relatively available instrument, like the hand, to feel in the chambers for vibrations and to report it in publication?

SUMMATION

Paul J. Kramer

Department of Botany
Duke University
Durham, North Carolina

INTRODUCTION

When I was requested to provide an "instant summary" at the
end of the workshop, I approached the task with mild optimism.
This later changed to despair as I realized that an ordinary
summary of the papers was not only difficult but would be a
tedious repetition of what had already been discussed. I therefore decided to concentrate on those topics which seemed to
illustrate some of the problems encountered in making environmental
measurements and on which I had definite personal opinions. Thus
my comments are highly personal and even opinionated, but they
are also influenced by nearly 25 years of controlled-environment
research.

The general objective of the workshop was to aid in development of a set of guidelines for the environmental measurements
which should be made by scientists doing research on plants in
controlled environments. There is urgent need for standardization
of instrumentation and of units of measurement so work done in
various laboratories can be compared. This work was started by a
committee of the American Society for Horticultural Science and
was also under study by a USDA committee. The USDA-SEA North

Central Region 101 Committee compiled a preliminary set of
guidelines which has provided an excellent outline for the organization of this workshop. In turn the discussion in the workshop
will be of great value to the committee in preparing the final
version of their guidelines.

In retrospect, two problems arose which could have been
avoided if foresight had been as good as hindsight. One problem
originated from the necessity of adopting SI (System International)
units. This difficulty could have been minimized by providing a
table giving the preferred SI units for various measurements.
Such a table might be included in the final version of the guidelines. The second problem involved recurrent confusion among the
discussants between the types of measurements required to characterize the environment in a growth chamber and those required by
investigators for their research. As stated earlier, the primary
objective of the workshop and of the committee on growth chamber
use was to promote standardization of methods of measuring
environmental factors so research done in different laboratories
can be compared. However, individual investigators often need
additional measurements and/or greater precision of measurements
than are required for comparisons of research studies and operation of controlled environment chambers.

This problem can be illustrated by reviewing some of the
topics discussed during the workshop.

RADIATION

Twenty years ago we usually disregarded warnings from physicists concerning the inadequacy of measurements of visible radiation with light meters because the only alternative was to measure
total radiation with a pyranometer which also was unsatisfactory.
Today we have photon flux meters which measure in the PAR
(photosynthetically active radiation) range and there seems to be
general agreement that results can best be reported in $\mu E\ m^{-2}\ s^{-1}$.

However, a minority favored use of moles of photons as more consistent with the use of SI units. Although most discussants
seemed to be satisfied with measurements in the PAR region,
others who work in photobiology reminded the workshop of the
importance of radiation in the ultraviolet and far-red regions
and proposed that measurements should be made over a range from
280 to 2800 or even 10,000 nanometers. Although necessary for
photobiologists, these measurements present serious difficulties
to most investigators, both because of the lack of inexpensive
instruments to make such measurements and because all of the
physiologically active wave bands have not been identified. It is
therefore both difficult to make and to interpret such detailed
measurements. It was agreed that in the absence of a complete
description of the intensity of radiation at various wavelengths
it would be helpful to provide a description of the kinds of
lamps used and the percentage of the total wattage used by each
type of lamp.

Measurement of intensity in moderately narrow wave bands,
perhaps 10 nm, would be useful to monitor changes in radiation
quality with aging of lamps and among different batches of lamps.
At present the high cost of such measurements makes them imprac-
ticable for most laboratories, but when suitable spectrophotometers
become available at a reasonable price these measurements should
be made. As one discussant stated, we must distinguish between
what is desirable and what is practicable. If we insist on
measurements that are impractical we will drive operators of some
growth chambers back to measuring illuminance in foot candles.

There was general agreement that radiation should be measured
at the beginning of each experiment and at least once every two
weeks afterward for long term experiments. The discussion concerning the location in a chamber at which measurements of
radiation should be made illustrates the difference between the
needs of the growth chamber operator and of the scientist who
uses it. If the operator wishes to maintain a uniform level of

radiation he must always measure radiation at some fixed distance
from the light source. However, the experimenter may wish to
measure radiation at the top of the plant canopy or at the level
of the leaves being studied which is quite different from the
measurement used to maintain uniform radiation in the chamber.

Duration of radiation was not discussed, but a suitable photoperiod must be selected for the species. This will depend on
whether vegetative growth or flowering is desired. Also, plants
subjected to a long photoperiod at full illumination receive much
more total radiation than those given a short photoperiod. One
way of minimizing this effect is by giving all plants a relatively short photoperiod of perhaps 12 hours at high intensity,
but producing long day effects by interrupting the dark period
with an hour of low intensity illumination. Care also must be
taken in greenhouses to avoid photoperiod effects from street
lights or even from night time auto traffic.

TEMPERATURE

Measurement of temperature produced less discussion than
measurement of radiation because there is unanimity concerning the
unit to be used and reasonable agreement on the use of shielded,
ventilated sensors. It was suggested that perhaps leaf temperature rather than air temperature should be controlled, but it
seems that the only generally practicable method of controlling
chamber temperatures is by controlling air temperature, because
different leaves differ in temperature with varying exposure,
radiation load, and air movement, If the investigator needs to
know the leaf temperature, as for calculation of vapor pressure
gradients, he can measure it. Differences between leaf and air
temperature are usually much smaller in growth chambers than in
full sun and therefore generally constitute a minor problem.

At one time considerable attention was given to thermo-
periodism (Went, 1953) and most growth chambers are routinely

operated at a night temperature 3 to 6°C lower than the day
temperature. However, it appears that such temperature changes
are unnecessary for some plants because they grow as well with
similar day and night temperatures as with different temperatures
if the constant temperature is favorable for growth (Friend and
Helson, 1976; Warrington et al., 1977). Another uncertainty concerns
the effects of small fluctuations in temperature. Evans (1963)
reported that temperature fluctuations of 2.5° about a mean
temperature of 22.5° resulted in more growth of tomatoes than a
constant temperature of 22.5°. Evans mentioned other examples of
effects on growth of short term fluctuations in environmental
conditions, and more research ought to be done in this area.

An important temperature problem in growth chamber research
is whether or not the soil should routinely be maintained at a
lower temperature than the shoots. It is well known that low
soil temperature reduces water uptake and sometimes affects root-
shoot ratios, and research on effects of cold soil on seed germination and seedling growth obviously requires independent control
of soil temperature. However, the available data appear to be
inadequate to evaluate the need for independent control of soil
and air temperature in routine research. It is clear, however,
that plants should not be irrigated with water at a temperature
very different from the soil temperature because cold water will
temporarily reduce absorption of water, and probably of nutrients,
by warm season species.

HUMIDITY

It has sometimes been argued that the atmospheric humidity is
not very important, but more recently it has been shown that many
kinds of plants, although not all, show increased growth at high
humidities. Also it has been shown that stomatal aperture of
some plants can be controlled by atmospheric humidity, independently of the bulk water status of the leaves. Thus, there are

adequate reasons to monitor and control humidity in controlled
environments. Humidification is best increased by use of steam,
but we find the use of a fine water spray satisfactory.

Use of the term "relative humidity" was questioned because it
can deceive the unwary. The rate of evaporation of water from a
moist surface is really controlled by the steepness of the vapor
pressure gradient from the leaf or other evaporating surface to
the air. At 20° the saturation vapor pressure of water is 23.3 mb,
at 30° 42.4 mb. Therefore, at 50% relative humidity the vapor
pressure gradient from leaf to air at 20° is 11.6 mb, but at 30°
it is 21.2 mb, so the rate of evaporation is nearly twice as
rapid at 30° as at 20° at the same relative humidity. In
experiments where plants are grown at various temperatures the
humidity should be adjusted so the vapor pressure deficit of the
air is similar, or at least varies no more than 5 mb among chambers. This will prevent atmospheric moisture from becoming an
unwanted variable. Hoffman recommends growing plants with a
vapor pressure deficit of 5 to 10 mb to minimize the effects of
atmospheric moisture. At 25° this would be approximately equal
to 70 to 80% relative humidity.

There was a lively discussion concerning the respective merits
of various kinds of humidity sensors, including wet bulb thermometers, dewpoint hygrometers, and a variety of other hygrometers.
Readers are referred to the discussion for more details. It
appears that a variety of sensors are being used successfully to
control growth chamber humidity, but most of them require recalibration at rather frequent intervals.

It was pointed out that growth chambers have several defi-
ciencies for water stress research. One is the inability to
obtain relative humidities or vapor pressure deficits as low as
those found out-of-doors in dry climates. Another is the frequent
failure to bring the humidity up to near saturation at night.
A third is the difficulty in imposing prolonged drying cycles
because of the small size of the containers. This is discussed
in the section on watering.

CARBON DIOXIDE

Although it has always been agreed that the concentration of CO_2 is important, it has only been generally realized in recent years that the concentration in growth chambers often is too low for optimum growth of plants. It seems that in order to maintain a constant concentration of CO_2 in growth chambers it is necessary to install equipment to monitor the concentration and inject CO_2 as required. Although measurement of CO_2 concentration with infrared gas analyzers is routine, users should remember that the instruments are pressure and temperature sensitive and errors can be caused by condensation of water in the gas line, use of gas lines permeable to CO_2, and desiccants that absorb CO_2. The most serious problem seems to be the unreliability of many of the gas mixtures sold for calibration of infrared gas analyzers. It is hoped that the National Bureau of Standards will develop standard gas mixtures suitable for this use. A relatively high air velocity also is necessary in the chambers to insure a uniform supply of CO_2 to all of the leaves.

It has been customary to express CO_2 concentration in parts per million (ppm) but SI terminology will require use of units such as micromoles per cubic meter of air. Some investigators use 350 ppm as their baseline concentration, but in urban areas the concentration is becoming so high that use of 400 ppm may become necessary. Another problem is the increase in CO_2 whenever people enter growth chambers or work in the area from which makeup air is drawn.

WATERING

Although the water supply of plants in growth chambers is the most easily controlled of all environmental factors it probably also is most often abused, and unrecognized water stress is of common occurrence. The investigator operates between the danger

of overwatering and creating root aeration problems and the danger
of underwatering and inhibiting growth by causing water stress.
Management of watering depends somewhat on the objectives of the
experiment and the nature of the medium used in the pots. If it
is desired to eliminate water stress the best approach probably
is to grow plants in gravel or a gravel-vermiculite mixture in
containers with drainage and irrigate the pots to the drip point
several times a day with a dilute nutrient solution such as half-
strength Hoagland solution. By applying solution to the drip point
and allowing the surplus to drain out each time, good aeration is
assured and salt accumulation is prevented. Applications can be
controlled by a time clock so the system is independent of human
errors, although it should be checked daily for mechanical failures.
The adequacy of watering can be tested by measurement of the leaf
water potential.

Chamber-grown plants are typically mesic in structure and
physiology because they are seldom subjected to water stress.
The only way to develop water stress is to withhold water until
stress develops. The rate at which stress develops depends on
the size of the plant, atmospheric conditions, the size of the
container, and the water holding capacity of the rooting medium.
Chamber-grown plants often fail to show characteristics of plants
stressed in the field, such as osmotic adjustment, because stress
develops too rapidly. They also are likely to show stomatal
closure at a higher water potential than field-grown plants.
These differences can be minimized by using larger pots containing a medium with a relatively high water holding capacity to
provide longer drying cycles. If transpiration rates are measured
by weighing, then water tight pots must be used or the pots
enclosed in plastic bags. The pot surface should be covered with
plastic to prevent loss by evaporation.

Water quality varies widely, but this can be controlled by
using deionized water as a solvent for nutrient solution and for
irrigation.

PRECISION

According to van Bavel, precision is a measure of our ability to quantify a process reproducibly, while accuracy refers to the agreement of data with reality. In many instances the readings observed on dials or print-outs are inaccurate because they vary from the real temperature, humidity, or CO_2 concentration by an unacceptable amount. Accuracy depends on calibration, but this requires reliable standards which are often lacking. Furthermore, calibration does not insure accuracy unless the sensor is used in the manner for which it was calibrated. For example, a pyranometer calibrated with a tungsten lamp or for sunlight will not accurately measure radiation from fluorescent lamps, and CO_2 in nitrogen does not give the same reading on an infrared gas analyzer as CO_2 in air. Temperature is the only environmental factor easily measured to less than 1% accuracy and, under chamber conditions, the errors in measuring other factors exceed 1% and often are 10% or even greater.

EXPERIMENTAL DESIGN

The proper design of experiments is extremely important if reliable information is to be obtained from growth chamber experiments. There are positional variations in the environmental conditions within chambers and there are variations among plants which make the need for replication and randomization as great in controlled environment chambers as in the field. Hammer expressed concern lest investigators use inappropriate computer programs and recommended that they consult with a statistician before starting their experiments. However, the writer has found that some statisticians have difficulty in dealing with the design of experiments for growth chambers.

PLANT CONDITIONS

Most of the discussion deals with instrumentation, but there
were numerous references to plant measurements and Berry and
Ulrich discussed interactions between plants and their environment. There were repeated warnings that the previous treatment of
plants may affect their subsequent behavior as experimental
material. Such occurrences as a period of mineral deficiency,
exposure to cold in the seedling stage, or prior exposure to
water stress may measurably affect subsequent behavior. Especially
disquieting was the observation that exposure of plants to wind or
mechanical vibration will significantly reduce growth. It is
believed that air movement sufficient to cause leaf flutter will
saturate the response to vibration, but this apparently has not
been fully demonstrated. It is possible that growth can be
reduced by vibration in a growth chamber, by moving the plants
from one chamber to another, or by the shaking involved in weighing
and measuring them. Furthermore, there are hazards from volatile
materials arising from paint or caulking compounds and other substances used in the laboratory or unknowingly admitted in makeup
air.

PLANT MEASUREMENTS

Although not treated as a special topic the nature of the
measurements to be made was referred to frequently. It seems
obvious that the kinds of measurements will vary with the objectives of the experiments and that we should decide what we wish to
measure before we start our experiments. Measurements can be
classified as physiological, biochemical, and morphological.
Physiological measurements such as measurement of CO_2 uptake,
transpiration, stomatal resistance, and plant water status in
terms of osmotic and water potential usually present no serious
difficulties. However, there were warnings concerning the

occurrence of oscillations in stomatal aperture (Barrs, 1971;
Levy and Kaufmann, 1976) which can cause large variations in
rates of photosynthesis and transpiration over periods of less
than an hour. Water and temperature stress also significantly
affect the relative amounts of various chemical fractions such as
the ratio of starch to sugar and saturated to unsaturated fatty
acids.

Investigations of the effects of temperature and other
environmental factors on growth are more meaningful if careful
morphological studies are made. Growth analysis involving
periodic measurement of roots, stems, and leaves yields much more
information than simple measurements of height and total dry
weight because they indicate which organs are being most affected
by a particular regime. The usefulness of growth analysis is
indicated by the Work of Patterson et al. (1979) who found that
in some species much of the effect of temperature on dry matter
production is brought about by its effects on leaf area expansion.

GENERAL CONCLUSIONS

The writer was impressed by the fact that although many important advances have been made in the field on controlled environment research, most of the basic problems under discussion 25
years ago are still with us, such as the kind and location of
sensors. However, there seems to be better appreciation of the
sources of error, both in measurement of the environment and in
respect to the past history of the plants used in the research.
Those who attended the workshop and readers of this volume certainly are more aware than previously of the limitations of our
measuring systems and the need for frequent recalibration of
instruments. They also are more aware of the effects of past
treatment on the future behavior of plants and the possibility
that sensitivity to wind and vibration, stomatal oscillations,

and uncontrolled variations in CO_2 content of the air can produce serious errors in our data. This knowledge should make future controlled environment research more reliable.

References to the various topics discussed in this summary can be found in other papers in this volume. However, a few additional papers are cited.

REFERENCES

Barrs, H. D. (1971). Cyclic variations in stomatal aperture, transpiration, and leaf water potential under constant environmental conditions. *Annu. Rev. Plant Physiol. 22,* 223-236.

Evans, L. T. (1963). Extrapolation from controlled environments to the field. *In* "Environmental Control of Plant Growth" (L. T. Evans, ed.), pp. 421-435. Academic Press, New York.

Friend, D. J. C., and Helson, V. A. (1976). Thermoperiodic effects on the growth and photosynthesis of wheat and other crop plants. *Bot. Gaz. 137,* 75-84.

Levy, Y., and Kaufmann, M. R. (1976). Cycling of leaf conductance in citrus exposed to natural and controlled environments. *Can. J. Bot. 54,* 2215-2218.

Patterson, D. T., Meyer, C. R., Flint, E. P., and Quimby, P. C., Jr. (1979). Temperature responses and potential distribution of itchgrass (*Rottboellia exaltata*) in the United States. *Weed Sci. 27,* 77-82.

Warrington, I. J., Peet, M., Patterson, D. T., Bunce, J., Hazlemore, R. M., and Hellmers, H. (1977). Growth and physiological responses of soybean under various thermoperiods. *Aust. J. Plant Physiol. 4,* 371-380.

Went, F. W. (1953). The effect of temperature on plant growth. *Annu. Rev. Plant Physiol. 4,* 347-362.

INDEX